Maria Alfaya

Botanical and seminal study of the bromeliad of the second Emperor of Brazil

Maria Alfaya

Botanical and seminal study of the bromeliad of the second Emperor of Brazil

Morphology and germination of seeds of Alcantarea nahoumil (Leme) J. R. GRANT. (Bromeliaceae).

ScienciaScripts

Imprint
Any brand names and product names mentioned in this book are subject to trademark, brand or patent protection and are trademarks or registered trademarks of their respective holders. The use of brand names, product names, common names, trade names, product descriptions etc. even without a particular marking in this work is in no way to be construed to mean that such names may be regarded as unrestricted in respect of trademark and brand protection legislation and could thus be used by anyone.

Cover image: www.ingimage.com

This book is a translation from the original published under ISBN 978-620-2-80813-2.

Publisher:
Sciencia Scripts
is a trademark of
International Book Market Service Ltd., member of OmniScriptum Publishing Group
17 Meldrum Street, Beau Bassin 71504, Mauritius
Printed at: see last page
ISBN: 978-620-3-36187-2

DEDICATION

To the Master of masters, the Lord Jesus Christ, because from Him, through Him and to Him are all things, and for having He enlightened my paths and made my dreams come true.

To my parents, Bernardino Ferreira and Maria Celina da Silva Ferreira (*in memorian*), for the efforts made while they lived, so that I could get here.

To my dear husband Antonio Alfaya, for his love, care and support.

To my beloved son, Nicola Ferreira Alfaya, for understanding my dream and the lack of lap in many moments...

Without them, I would not be realizing this dream

THANKS

To God, for being absolutely everything in my life and for having guided my steps at every moment of this work.

To IF Baiano and UFRB for the opportunity to take the course.

To CAPES, for the financial support.

To the professors, Dr. Maria Angélica Pereira Carvalho Costa and Dr. Edson Ferreira Duarte, for their orientation, teachings, dedication, friendship, and mainly for their patience.

To my husband Antonio Alfaya, for his love, dedication and care.

To my sisters, Maria José Ferreira Costa and Iraci Ferreira da Mota, for their valuable help and incentive.

À all my family members, for always rooting for my success.

À Prof. Dr. Ana Cristina Vello Loyola Dantas, Course Coordinator.

To Prof. Dr. Maria Cristina Fermino Soares, Coordinator of the Master's Minter. To the professors of this post-graduation program, especially to professor Dr. Carlos Alberto da Silva Ledo for his collaboration in the statistical analyses.

To the doctoral researcher Daniela Sousa Hansen, for the valuable collaboration in locating the plants and collecting the seeds.

À Dr. Maria das Graças Lapa Wanderley, from the Botanical Institute of São Paulo, for her contribution in the taxonomic identification of the species.

To the students, Fábio Ribeiro Garcia, Érika Ribeiro Souza, Camila Gonzaga de Jesus, Jailson de Souza Pereira, Jaylson Araújo dos Santos, for their collaboration in the laboratory activities.

To professor Julian Perez Pizarro, for his help in translating the texts. To all course colleagues and friends for their untiring help, companionship and friendship, which contributed so much to minimize the difficulties. To my friends at IF Baiano, Rozenaide Pires Lima, Maria Aparecida Farias de Vasconcellos, Admilson Santana, Ana Rute da Silva, Ana Rita Motas. And to the teachers and directors, for the friendship and incentive.

Thank you very much

SUMMARY

SEED MORPHOLOGY AND SEED GENERMINATION OF *Alcantarea nahoumii (*Leme) J. R. GRANT. (BROMELIACEAE).

Author: Maria Cristina Ferreira Alfaya
Advisor: Prof. Maria Angélica Pereira de Carvalho Costa
Co-Supervisor: Prof. Edson Ferreira Duarte

Alcantarea nahoumii is a rupicolous bromeliad, endemic to rocky outcrops in the Serra da Jibóia, Bahia, Brazil, with high ornamental potential. The objective was to describe the morphology of the fruits, seeds and post-seed development, to evaluate substrates and temperatures on germination and vigor of seedlings, in order to increase the understanding of reproductive strategies and to enable its conservation. Fruit, seed, and seedling morphology was described and illustrated. Germination was tested at 25ºC in BOD with 4 repetitions of 100 seeds in sand, paper, Plantmax® and vermiculite, which were physically characterized. Germination on paper in the dark was also evaluated. The effect of temperature was evaluated on germinated seeds at 20, 25, 30 and 35 °C. Seedling survival was evaluated at the end of the experiments by the measurement of shoot and root length and total fresh mass. The first count and the final count were determined. The fruits are septate capsules, with spontaneous dehiscence. The seeds are filiform, small and light, with the presence of proximal and distal papus. Germination is cryptoepigeial, beginning at 10 days. Seedlings show slow development. Germination occurs preferentially in the presence of light, and the substrates have little effect on this, affecting the vigor of the seedlings. The highest seedling growth was provided by Plantmax®, after 60 days. Germination at 35 °C limited the germination process and the development of the seedlings. The morphology of the seeds enables their dispersion by wind, but the slow development of the seedlings may reduce recruitment in their natural environment. Substrates with density, total porosity, and intermediate water holding capacity can be used in germination tests at temperatures of 25°C. The first count can be done at 18 days and the final count at 54 days.

Keywords: substrate, growth, seedlings, temperature, vi gor.

INTRODUCTION

The Bromeliaceae family consists of three subfamilies, Bromelioideae, Pitcairnioideae and Tillandsioideae (LEME and MARIGO, 1993). The latter contains the genus *Alcantarea*, native to Brazil and has about 15 species. Its name is in honor of Dom Pedro de Alcantara (1840-1889), the second Emperor of Brazil. Such species are rupicolous, growing naturally on rocky outcrops or shallow stony soils, being exposed to high light, with great plasticity to temperature variations, being considered tolerant to cold (ANDREAS, 2006).

The economic importance of bromeliads is related mainly as ornamental plants, being cultivated and used in interior decoration and landscape projects (MENDONÇA, 2002). Due to the great demand for bromeliads with ornamental value, the extractivism has intensified in recent years, placing some species with a greater degree of threat (MOREIRA et al., 2006).

Due to the great wealth of flora in Brazil there is still a lack of research that provides knowledge of native species, especially in seed germination and post-seeding stages of plant development, which can serve as a reference and subsidy for programs of recovery and management of natural areas. Thus, the expansion of knowledge about the behavior of native species, in their post-seeding stages, contributes to their use can be intensified.

According to the Program for the Protection of Endangered Species in the Brazilian Atlantic Forest, of the Biodivers itas Foundation, *Alcantarea nahoumii* is a bromeliad species that is in the category of "Species vulnerable to extinction" (BIODIVERSITY, 2010) and is on the list of Data Deficient Species of the Brazilian Flora (BRASIL, 2008). Therefore, the development of conservation strategies is fundamental for its protection against extinction.

Research on seedling development and morphology has been carried out with different approaches, either for the recognition and identification of the initial stages of development of species from a certain region or ecosystem, since seedling emergence and establishment are critical stages in the plant life cycle, or as part of morpho-anatomical research of certain species or systematic groupings (MORAES and PAOLI, 1999; ROSA et al, 2005; OLIVEIRA, 2001; MELO and VARELA, 2006; MELO et al., 2004).

Other works emphasize the area of technology, analysis and conservation of seeds (ANDRADE et al., 2003). However, information on seed germination, substrate types, growth, requirements and tolerances of native plants is scarce, as already observed by Labouriau (1983).

A number of professionals have dedicated themselves to the determination of standards to be adopted for the quality detection of seed lots of native species, with a view to their storage and utilization, and their conservation in germplasm banks (CARMONA, 1999).

The present work aimed to describe the morphology of fruits, seeds and post-seed development, and to evaluate substrates and temperatures on seed germination and seedling vigor of *Alcantarea naohumii,* in order to increase the understanding of reproductive strategies and enable its conservation.

Atlantic Rainforest Biome

The Atlantic Forest is a biome present in most of the Brazilian territory, also covering part of the territory of Paraguay and Argentina. The original area was 1,290,692.46 km², 15% of the Brazilian territory, however, due to deforestation and disorderly occupation, today the remnant is 95,000 km², 7.3% of the original area. Today most of the coastal area that used to be covered by the Atlantic Forest is occupied by large farms, pastures and agriculture. However, patches of the forest still remain in the Serra do Mar and Serra da Mantiqueira in southeastern Brazil. Today it covers 3.409 municipalities in 17 Brazilian states (RS, SC, PR, SP, GO, MS, RJ, MG, ES, BA, AL, SE, PB, PE, RN, CE, PI) (SOS Foundation and INPE, 2000).

This biome has clearly defined vegetation layers. The tops of the tall trees form the canopy and reach heights of 30 to 60 meters. The tree trunks are usually smooth and branch out only high up to form the canopy. The crowns of the taller trees touch each other, forming a mass of leaves and branches that block the sun's path. In the lower part, shrubs and small trees grow, such as bamboos, giant ferns, and lichens that tolerate less light, forming the so-called understory. In the taller trees as well as in the lower ones you can find various other species, such as different types of lianas, bromeliads, orchids, and tendrils. The forest floor is covered by the undergrowth. This floor is protected by the leaves and other vegetation that fall from the trees throughout the year, which serves as food for many insects, other animals, and especially the fungi, which are mainly responsible for the process of decompostion of the forest. Thus, the forest feeds on itself (MARTINS, 2006).

The climate, in the region encompassed by the Atlantic pluvial forests, has two seasons, defined mainly by the rainfall regime, although it is latitudinally quite variable. While in the Brazilian North the average annual temperatures vary around 24 ºC, in the Southeast and South the annual averages are lower and the temperature can occasionally reach 6 ºC (MARTINS, 2006).

Forest soils are usually poor in mi nerals and their nature is granitic or gneissic. Most of the minerals are contained in the plants rather than in the soil. As there is a lot of litter in the soil that gives rise to abundant humus, there are microorganisms of various kinds that decompose the organic matter that is incorporated into the soil. These minerals, once released by the decomposition of leaves and other debris, are readily reabsorbed by the large number of existing roots, returning to the soil when the plants or their parts (branches, leaves, flowers, fruits and seeds) fall. Thus, the plant-soil cycle is closed, which explains the maintenance of exuberant forests, in soils that are not always fertile, sometimes poor (NOGUEIRA, 2008).

Floristic inventories in several parts of the Atlantic domain have pointed Bromeliaceae among the families with greatest richness and diversity, both generic and specific (BARROS 1991; MARQUES, 1997; LIMA e GUEDES-BRUNI, 1997; ARAÚJO, 2000; COSTA e DIAS, 2001; MAME DE et al., 2001; ASSIS et al., 2004; AMORIM et al., 2005; MARTINELLI, 2006). Among the species of Bromeliaceae recorded in the Atlantic Forest about 40% are under some category of threat (there are species in more than one category). Furthermore, it is likely that this number is underestimated due to the little knowledge of the real conservation status of populations in natural environments, as well as the reduced number of specimens deposited in herbaria, which compromises the evaluation of the geographical distribution of taxa (MARTINELLI et al.,2008).

Family Bromeliaceae

The Bromeliaceae family has 3172 species distributed in 58 genera (LUTHER, 2008), divided into the subfamilies Bromelioideae, Pitcairnioideae, Tillandsioideae, according to the floral and morphological characteristics of the fruits and seeds (SMITH and DOWNS, 1974; PAULA and SILVA, 2004).

The species classified as fast belong to the Bromelioideae, except *Dyckia pseudococcinea* L. B. Smith (Pitcairnioideae); the intermediate and slow species belong to the Tillandsioideae and *Pitcairnioideae*, respectively, (PEREIRA et al, 2008), which occurs in western Africa (SMITH and DOWNS, 1974).

Bromeliads are characterized by epiphytic plants, terrestrial and rupicolous, with generally short stems covered by leaf sheaths. The leaves are alternate, polystic or dystic, in general forming a rosette around the stem. The leaf surface is covered with specialized trichomes (leaf scales) and the margins can be entire, serrated or spinescent, with usually enlarged sheaths. The inflorescences are usually racemose, simple or branched. The scape can be long to nearly sessile, and are often covered with foliaceous or colorful bracts. The flowers are usually conspicuous contrasting with the bracts, and the fruits are capsules or berries with seeds that can be appendiculate or not (ROCHA, 2002).

It is considered an important source of nutrient cycling for forests with poor soils (OLIVEIRA, 2004). Several animals use bromeliads for foraging, reproduction and refuge against predators (COGLIATTI-CARVALHO et al., 2001), and some species have been described as bioindicators of atmospheric pollution (BRIGHIGNA et al., 2003; ELIAS et al., 2006).

Bromeliad species occur in tropical and subtropical latitudes of the Americas between parallels 37°N and 44°S under widely varying conditions of altitude, temperature, and relative humidity (WENDT, 1999).

Approximately two-thirds of bromeliads are CAM plants and occupy arid sites or are epiphytes. Many species live in total or partial shade and others are fully exposed to the sun (MA RTIN 1994).

CAM metabolism (crassulacean acid metabolism) is a photosynthetic CO2 concentrating mechanism and was possibly selected in response to the aridity of terrestrial environments (KEELEY 1998). In these plants, the fixed CO2 is accumulated in the vacuoles as malate, contributing to cell acidity. During the day, the stomas close, and the CO2, used in the C3 cycle, is supplied by the decarboxylation of malate. During the night the starch that was produced and accumulated starts to be hydrolyzed for the production of PEP, accumulating during the day as a product of photosynthesis and the decarboxylation of malate (MAJEROWICZ, 2008).

The bromeliads present a great variability of form, being generally very characteristic and ornamental plants. According to Benzing (2000), the different habitats and especially the nature of the substrate influence the appearance of the plant, which can vary widely in size and coloration of the leaves, as well as in the morphology of the flowers.

The bromeliads became known in Brazil in the years 1940 to 1950, when they were the target of an extermination attempt by the government, aiming to control the mosquito that transmitted malaria. All specimens that could be reached from the trees were destroyed, drastically reducing the population levels (REBELO, 2004). In 2000, the bromeliads were again the target of unjust persecution and mass destruction with the advent of the dengue fever epidemic in Brazil. Hundreds of thousands of bromeliads were mistakenly destroyed in order to avoid the proliferation of the mosquito that transmits the disease in the water accumulated in their phytotelm, but the dengue mosquito vector does not use the water stored in the bromeliads under natural conditions, because it reproduces in clean water and the bromeliads always have organic matter in decomposition (CARVALHO, 2002).

Bromeliads, initially marginalized, were later used and sought after as an ornamental plant, due to the repercussion of their use by the internationally renowned landscape designer Roberto Burle Max (ENGLERT, 2000), thus beginning the cycle of bromeliad extraction. In Brazil this extraction is generally associated with communities living on the margins of the Atlantic Forest.

In large cities, in many ornamental gardens where these plants are used, it is common to use plants extracted from the forests. The cycle of extraction of plants from nature, especially bromeliads, for commercialization occurs intensely nowadays.

Anacleto (2005) verified that in some locations, such as the south of the metropolitan region of Curitiba and the coast of Paraná, bromeliads suffer great extractive pressure, originated by poor families living there and also by ornamental plant traders. In this region the predominant vegetation is the Atlantic Forest, which is an ecosystem considered one of the largest centers of diversity of the Bromeliaceae family. It is noteworthy that several species occur in very restricted and reduced areas, for this reason the bromeliad population is considered endemic, a fact that considerably increases the risk of extinction, either by deforestation or by indiscriminate extractivism (BENZING, 1980; NUNES and FORZZA, 1999), and in both cases there may be a loss of genetic diversity caused by man (LEME, 1997).

All efforts should be directed to the conservation of its remaining natural populations, because changes in this system can endanger the survival of the species or compromise the habitat that survives associated with these plants (NAHOUM, 1994), since according to Tardivo (2002) Paraná has the most conserved portion of Atlantic Forest in Brazil, and still little is known about the bromeliological flora of this state.

It is known that large quantities of these plants are collected from the forests or forest reserves to be sold in the domestic market or for export. The awareness about nature conservation and the commercial production of bromeliads can ease this situation (ANDRADE e DEMATTÊ, 1999).

The production of bromeliads on a commercial scale is a profitable activity. The quality of the plants that come from commercial production is superior to those that come from criminal extractivism. Furthermore, the reduced costs place the plants cultivated in nurseries at a great advantage in the market (MELO, 1996).

Bromeliads can be propagated sexually or asexually. The sexual process involves the formation of seeds, from which large quantities of seedlings can be obtained (PAULA, 20 05). However, the sexual propagation of bromeliads is time-consuming, because just the maturation of the seeds can take up to a year after pollination, depending on the species. One of the aspects observed in the cultivation of some species is that despite the large number of seeds produced, the seedlings obtained by sowing have slow growth and development (STRINGHET A et al., 2005).

In the asexual propagation, or vegetative, buds are formed from the mother plant, which can leave the base of the plant through stolons or rhizomes, or from inside the rosette itself. The formation of stolons is characteristic for some species (MOREIRA et al., 2006).

The seeds of Tillandsioideae are small, light and have morphological adaptations that increase the surface to volume ratio, reducing the speed of fall (PAULA and SILVA, 2004). Generally, they use air currents for transportation in dry periods of the year, facilitating the dispersion of these seeds within the crevices of the rocks, where they find ideal germination conditions. The feathery appendages help fix the seeds on trunks and tree barks, ensuring the success of their dispersion (VAN DER PIJL, 1982; BENZING, 2000; SCATENA et al., 2006).

According to Moreira et al. (2006) sexual reproduction is frequent in representatives of the Tillandsioideae subfamily, where the seeds can germinate very close to the mother plant or be dispersed over long distances .

Seed dispersal represents an extremely important phase of the reproductive cycle of plants, being crucial in the regeneration of populations and natural communities (JANZEN, 1988). Seeds often have adaptations that facilitate dispersal, which are quite evident in their morphology. Knowledge of seed morphology allows us to relate them to their dispersal agents (ALMEIDA-CORTEZ, 2004) and to differentiate taxonomic groups (SCATENA et al., 2006).

Characteristics of *Alcantarea nahoumii* (LEME) J. R. GRANT.

Alcantarea nahoumii (LEME) R. J. Grant belongs to the Tillandsioideae subfamily, has rupicolous habit, growing naturally on rocky outcrops (Figure 1a) or shallow stony soils, being exposed to high luminosity, and can reach up to 2.4 meters in height, being considered a species of intermediate growth (PEREIRA et al., 2008).

Figure 1. *Alcantarea nahoumii* (Leme) J. R. Grant (Bromeliaceae). a. Plant in natural conditions in Serra da Jibóia-BA. b. Detail of the inflorescence.

The species was described by Grant (1995) as a rupicolous, flowering plant about 100 centimeters tall, has suberect, densely rosulate leaves forming a rosette, broadly funneliform in the proximal region. Sheath broadly elliptic, about 10 centimeters wide, pale in color, with brown trichomes, densely lepidote on both sides. Leaf lamina sublinear, 35 centimeters long and 7 centimeters wide, not narrow in the prox imal region, with short distal region, acuminate and curved, partially glabrous and green on both faces.

The floral scape is about 70 centimeters long, with a diameter varying from 0,8 to 1,3 centimeters, rigid, erect, grooved, red and glabrous. The scapular bracts are suberect, becoming narrow, with recurved distal region, of reddish color, with the proximal region hiding the scape, of subfoliaceous consistency, bigger than the internodes, covering it from 3 to 6 centimeters above, being suborbicular with the di stal region pointed or sub acute and apiculate.

The inflorescence is short-paniculate, bipinnate, subdescending, about 20 to 30 centimeters long. The primary bracts are orbiculate, obtuse and short-apiculate, about 3 centimeters long and 3 centimeters wide, reddish, similar in appearance to the sterile bracts of the secondary rachises.

The primary raceme is geniculate, grooved and glabrous, greenish with 5 millimeters in diameter. Orbiculate floral bract about 3 centimeters long, with obtuse distal region, concave, curved close to the proximal region, subcoreaceous, ecarinate and reddish, inc onspicuously lepidate in the internal face, not imbricate after the anthesis, covering the sepals, slightly short, equivalent to the size of the sepals.

The secondary spike is about 7 centimeters wide (excluding the petals), underexpanded, unilateral and arched, with acropetal blooms; in lateral view, they measure 4 to 11 centimeters and 10 to 15 centimeters long, composed of 13 to 16 pedunculated flowers about 3 centimeters long and 0.5 centimeters in diameter; it is compact, reaching up to 20 centimeters long with the terminal flower.

The flowers are alternate dystic (Figure 1b), expa ndent in the proximal region, about 3.5 centimeters long (excluding the petals). The pedicels are robust, slightly coplanate, about 6 millimeters long and 8 millimeters in diameter. The sepals are slightly asymmetrical, obovate, rounded, about 30 millimeters long and 17 millimeters wide, glabrous, ecarinate, yellowish, dialsepalous. The petals are observed only in the young flower or in the depauperate follow-ups, the distal region is narrow, obtuse, becoming biobovate, fixed in the proximal region; the stamens are about 70 millimeters long, with obtuse proximal and distal regions, located in the proximal region.

A. nahoumii is native to the Serra da Jibóia, occurring also in other regions in the State of Bahia, Area of Tropical Forest al and Campos de Altitude, which approach 800 meters. (MARTINELLI et al, 2008).

The Serra da Jibóia has an extension of 6 km and is located in the municipality of Santa Terezinha - Bahia, at a latitude of 12°51'S, longitude 39°28'W, altitude of 750-800 m, rainfall should reach at least 1,100 mm annually. At the top of the mountain is a gneissic-granite outcrop, over which is developed a rupestrian field vegetation. The fact of being inserted in the semi-arid region of Bahia makes the hygrophilous forest of Serra da Jibóia, a unique place, since it is isolated from other forest fragments of the coast by the surrounding caatinga (QUEIROZ et al., 1996).

Substrates for growing bromeliads

There are different materials available on the market and in nature that can be used as substrates, among them: sand, soil, sawdust, carbonized rice husk, vermiculite, Plantmax®. Adequate conditions for germination and/or emergence and development of the root system depend on the absence of pathogens and weeds, adequate pH, texture and structure of the substrate (SILVA et al., 2001).

For Kampf (1992), the cultivation of epiphytic bromeliads requires substrates with low density and high permeability and aeration; the presence of organic matter in the culture medium improves these characteristics. The main component should have a high C:N ratio, that is, be difficult to decompose, allowing good aeration and drainage of the substrate (PAULA, 2005).

The first component used as substrate was soil, and it is still used today, especially for large perennial plants. Sand and burlap were used to grow epiphytic bromeliads in pots at the end of the 19th century, according to Kampf (1992). After the Second World War, in Europe, research into the use of peat as a growing medium for potted plants, which had been interrupted during the war, resumed. Later, other materials were incorporated to the productive sector, such as perlite in Spain and bark in the United States and Europe. In Brazil, the main substrate components are materials of organic origin such as peat, pine bark and coconut fiber (BATAGLIA and FURLANI, 2004).

The density of the substrate is the ratio between the mass of dry material and the volume, expressed in grams per cubic centimeter. The density is correlated with the characteristics, total porosity, available water and aeration space (SALVADOR, 2000, FERMINO, 2002). For most plants the value is between 0.3 and 0.4 g. cm^{-3} (BALLESTER and OLMOS, 1992). The stability of the structure and the granulation are decisive factors, because the macropores directly affect aeration and moisture (BARBOSA, 2007).

The determination of total porosity can be done by the water retention curve, and corresponds to the moisture content at zero pressure, that is, when it is at saturation point. Its estimation is an essential precondition for the evaluation of the space occupied by air and the capacity to retain and release water to the plants (WALLER and HARRINSON, 1991). The ideal value indicated by Verdonck and Gabriels (1988) is 85%.

The available water is the difference between the amount of water in the substrate after drainage, corresponding to the water retained between 10 and 100 cm in tension. The optimum value for available water ranges between 20 and 30%. High values cause hypoxia of the roots, while low values mean the need for frequent irrigation (BALLESTER-OLMOS, 1992).

Seed Germination

Germination is the process that usually starts with the absorption of water followed by an increase in the metabolic activity of the seed, promoting the intra-seminal growth of the embryo, culminating with the protrusion of the primary root or plumule (LABOURIAU, 1983). Bewley and Bla ck (1986) reported that in some seeds of the families Bromeliaceae, Palmae, Chenopodiaceae, Onagraceae, Saxifragaceae and Typhaceae, the hypocotyl is the first structure to emerge. More specifically in the Bromeliaceae Smith and Downs (1974) report that germination occurs with the elongation of the lower portion of the cotyledon that pushes the radicle out of the seed, the plumule remains enveloped by a cotyledonary sheath, while the upper portion of the cotyledon remains internally in the seed absorbing the reserves of the endosperm, keeping the seed attached to the seedling for some time. The primary root ceases its growth when it reaches about 0.50 cm to 1.00 cm in length (DUARTE, 2007).

Germination is affected by internal factors such as seed viability and longevity, and also by external factors such as water availability, oxygen availability, adequate temperature and light (DUARTE, 2007). These factors affect the seeds at a certain time or throughout germination. Under natural conditions, germination regulatory mechanisms can be advantageous to the species, preventing their germination in unfavorable or unstable conditions, thus, the survival and germination give the species more chances to establish themselves in competition with others occupying the same habitat (MAYER and POLJAKOFF- MAYBER, 1963).

There is a consensus among seed physiologists and technologists that the germination process starts with soaking (MARCOS FILHO, 2005). Bewley and Black (1986) report that water absorption occurs in three phases: the first phase of soaking is due to the action of the malic potentials, occurring in seeds, regardless of being viable dormant or not. In the next phase, the water continues penetrating the tissues, at a lower velocity. In the third phase there is a rapid increase in water absorption, marked by the protrusion of the primary root.

The germination process meets arbitrary criteria that aim to determine the beginning and the end of germination. Some authors consider that the process begins from the intensification of respiration of the seeds, however, there are difficulties in observing this phenomenon. The absorption of water is another criterion, but dead seeds may present soaking of hydrophilic constituents, leading even to protrusion of the radicle of dead embryos, and therefore the adoption of this criterion is inadequate. So, to consider a seed as germinated, it is necessary that it presents some sign of life, such as geotropism of the primary root or the opening of the plumule hook (Papilionoideae). In certification work, the germination criterion should also be adopted for seeds that originate normal seedlings (LABOURIAU,1983).

The *Association of Official Seed Analysts of the* United States of America recommends that seeds that show emergence and development from the seed embryo of essential structures that, for the species in question, indicate its ability to produce normal seedlings under favorable conditions, be considered germinated (AOSA, 19 83).

Effect of temperature on seed germination and seedling vigor

Seed germination is composed of a series of morphological events that result in the transformation of the embryo into a seedling. The evaluation of seed germination is done by the germination test, conducted in a laboratory under controlled conditions and using standardized methods that aim, mainly, to assess the value of seeds for sowing and to compare the quality of different lots, serving as a basis for the commercialization of seeds (MARCOS FILHO et al., 1987; NOVEMBRE, 1994).

According to Taiz and Zeiger (2004), when tropical and subtropical species are subjected to temperatures between 15 °C and 10 °C, they grow more slowly.

Very high temperatures cause stress or even loss of viability. High temperatures decrease the supply of free amino acids, protein synthesis, RNA synthesis, and anabolic reactions. In general, high temperatures denature proteins, alter membrane permeability, and cause loss of material, while low temperatures slow metabolic rates to the point where the essential pathways for the initiation of germination can no longer operate (CASTRO and VIEIRA, 2001).

BIBLIOGRAPHIC REFERENCES

ALMEIDA-CORTEZ, J. S. Dispersão de sementes. In: FE RREIRA, A. G.; BORGHETTI, F. (eds.). **Germination: do básico ao aplicado**. Porto Alegre, Artmed, 2004. p. 225-235.

AMORIM, A. M. A.; FIASCHI, P.; JARDIM, J. G.; THOMAS, W. W.; CLIFTON, B. C. & CARVALHO, A. M. V. The vascular plants of a forest fragment in Southern Bahia, Brazil. **Sida** v. 21, n. 3, p. 1727-1757. 2005.

ANACLETO, A. **Seed germination and shoot development of Aechmea nudicaulis (L) GRISEB. (Bromeliaceae) Subsidies to production and extractivism.** Dissertation (Master in Agronomy), UFPR, Curitiba, PR. 2005. 74p.

ANDRADE. F. S. A.; DEMATTÊ, M. E. S. P. Estudo sobr e produção e comercialização de Bromélias nas regiões Sul e Sude ste do Brasil. **Revista Brasileira de Horticultura Ornamental,** Campinas, v. 5, n. 2, p. 97-110, 1999.

ANDRADE, A. C. S.; CUNHA, R.; SOUZA, A. F.; REIS, R. B.; ALMEIDA, K. J. Physiological and morphological aspects of seed viability of a neotropical savannah tree, *Eugenia dysenterica* DC. **Seed Science & Technology, n.** 3, p. 125-137, 2003.

ANDREAS, K. Growing Alcantarea Species: Illustrating Terrie Bert's Article. **Newsletter of the Bromeliad Society of Central Florida,** v. 32, n. 2, 2006. Disponível em: < http://fcbs.org/newsletters/BSCF/022006.pdf>. Acesso em: 16/ dez. 2008

AOSA, Association of Official Seed Analysts. **Seed vigor testing handbook**. AOSA. 1983. 88p. (Contribution n° 32 To the handboo k on Seed testing).

ARAUJO, D. S. D. **Análise florística e fitogeográfica das restingas odEstado do Rio de Janeiro**. Tese. (Doctorate in Ecology), Federal University of Rio de Janeiro, Rio de Janeiro, 2000. 176p.

ASSIS, A. M.; THOMAS, L. D. and PEREIRA, O. J. Floristics of a stretch of restinga forest in the municipality of Guarapari, Espírito Santo, Brazil. **Acta Botânica Brasilica.** v.18, n.1, p.191-201. 2004.

BALLESTER-OLMOS, J. F. **Substrate for the cultivation of ornamental plants**. Valencia: Instituto Valenciano de investigaciones Agrárias, 1992. 44p.

BARBOSA, G. C. V. **Substrate and flowering inducers in ornamental bromeliads.** Dissertation (Master in Phytotechnology). Universidade Federal de Viçosa. Viçosa, 2007. 77p.

BARROS, F. de; MELO, M. M. R. F. de; CHIEA, S. A. C.; KIRIZAWA, M.; WANDERLEY, M. das G. L. and JUNG-MENDAÇOLLI, S. L. Ca racterização geral da vegetação e lista das espécies ocorrentes. *In*: **Flora Fanerogâmica da Ilha do Cardoso** (MELO, M. M. R. F.; BARROS, F.; WANDERLEY, M. G. L.; KIRIZAWA, M.; JUNG-MENDAÇOLLI, S. L. E CHIEA, S. A. C. eds.). Institute of Botany, v.1, São Paulo, 1991. p. 1-184.

BATAGLIA, O. C.; FURLANI, P. R. Nutrição mineral e adubação para cultivos em substratos com atividade química. In: BARBOSA, J. G.; MARTINEZ, H. E. P.; PEDROSA, M. W.; SEDIYAMA, M. A. N. **Nutrição e adubação de plantas cultivadas em substrato**. Viçosa, MG: UFV, 2004. p. 106 -128.

BENZING, D. H, **The biology of the bromeliads**. Mad River Press, California, EUA, 1980. 305 p.

BENZING, D. H. **Bromeliaceae**: profile an adaptative radiation. Cambridge University Press. 2000. 690p.

BEWLEY, D.; BLACK, M. **Seeds:** physiology of development and germination. New York: Plenum Press, 1986. 367p.

BIODIVERSITY. **Consultation on the Revision of the List of Brazilian Flora Threatened with Extinction.** Available at <http://www.biodiversitas.org.br/florabr/grupo3fim.asp>.

Accessed on 09/May 2010.

BRAZIL. Ministry of Environment. **Official List of Species of Brazilian Flora Threatened with Extinction.** Normative Instruction Nº 6, of September 23, 2008. 55p.

BRIGHIGNA, L.; PAPINI A.; MOSTI S.; CORNIA A.; BOCCHINI P.; GALLETTI G. The use of tropical bromeliads (*Tillandsia spp*) for monitoring atmospheric pollution in the town of Florence, Italy. **Revista de Biologia Tropical**, v.50, n. 2, p. 577-584, 2003.

CARVALHO, L. F. N. **O cultivo da bromélia.** São Paulo: TJV Publishing, 2002. 32 p.

CASTRO, P. R. C.; VIEIRA, E. L. **Aplicações de Reguladores Vegetais na Agricultura Tropical.** Guaíba: Agropecuária, 2001. 132 p.

COGLIATTI-CARVALHO, L.; NUNES-FREITAS, A. F.; ROCHA, C. F. D. and VAN SLUYS, M. Variation in the structure and composition of Bromeliaceae in five vegetation zones in the Restinga de Jurubatiba National Park, Macaé, RJ. **Revista Brasileira de Botânica** , v. 24, n.1, p. 1-9, 2001.

COSTA; A. F.; DIAS, I. C. A. (orgs.). **Flora of the Restinga de Jurubatiba National Park and surroundings, RJ: listing, floristics and phytogeography (Angiosperms, Pteridophytes and Continental Algae)**. Museu Nacional / UFRJ, Rio de Janeiro, 2001. 200p.

DUARTE, E. F. **Caracterização, qualidade fisiológica de sementes e crescimento inicial de *Dyckia goerhringii* Gross & Rauh, bromélia nativa do cerrado.** Thesis (Doctorate in Agronomy). Universidade Federal de Goiás. Goiás, 2007. 200p.

ELIAS C.; FERNANDES E. A. N.; FRANÇA E. J.; BACCHI M. A. Selection of chemical element accumulating epiphytes in the Atlantic Forest. **Biota Neotropica**, São Paulo, v. 6, n. 1, p. 1-9, 2006.

ENGLERT, S. I. **Orquídeas e Bromélias, manual prático de cultivo.** Editora Agropecuária, Guaíba-RS, 2000. 92p.

FERMINO, M. H. O uso da análise física na avaliação da qualidade de componentes e substratos. *In:* ENCONTRO NACIONAL DE SUBSTRATOS PARA PLANTAS, 3. Campinas. ***Proceedings...*** Campinas: IAC, p. 29-37, 2002.

FUNDAÇÃO SOS MATA ATLÂNTICA; NATIONAL INSTITUTE FOR SPACE RESEARCH. **Atlas of Forest Remnants and Associated Ecosystems in the Atlantic Forest Domain.** São Paulo, 2000. Available at:
< http://www.sosmatatlantica.org.br>. Accessed 26/May. 2009.

JANZEN, D.H. Management of habitat fragments in a tropical dry forest: growth. **Annals of the Missouri Botanical Garden** v.75, p.105-116, 1988.

KAMPF, A. N. Substrato para floricultura In: **Simpósio Brasileiro de Floricultura e Plantas Ornamentais**, Maringá, 1992, Manual de floricultura. Maringá, S BFPO, p.36-43. 1992.

KEELEY, J. E. CAM Photosynthesis in Submerged Aquatic Plants. **The Botanical Review** n. 64, p.121-175, 1998.

LABOURIAL, L. G. **The Germination of Seeds**. Washington: General Secretariat of the Organization of American States, 1983. 173p.

LEME, E. M. C.; MARIGO, L. C. *Bromeliads in nature.* **Marigo Comunicação Visual Ltda**, Rio de Janeiro, 1993. 183p.

LEME, E. M. C. **Canistrum, Bromélias da Mata Atlântica.** Ed. Salamandra, Rio de Janeiro, 1997.107p.

LIMA, H. C. and GUEDES-BRUNI, R. R. Diversidade de plantas vasculares na Reserva Ecológica de Macaé de Cima. In: LIMA, H. C. and GUEDES-BRUNI, R. R. (orgs.). **Serra de Macaé de Cima: Floristic Diversity and Conservation in Mata Atlântica.** Instituto de Pesquisas Jardim Botânico do Rio de Janeiro, Rio de Janeiro. 1997. p. 29-39.

LUTHER. H. E. **An alfabetical listo f Bromeliad binomials**. 11º ed. The Marie Selby Botanical Gardens Sarasota, Flórida, USA. The Bromeliad Society International, 2008. 109 p.

MAJEROWICZ, N. Photosynthesis. In: KERBAUY, G. B. **Fisiologia Vegetal** - 2ª. ed. Guanabara Koogan, Rio de Janeiro, 2008. p. 82-133.

MAMEDE, M. C. H.; CORDEIRO, I. and ROSSI, L. Flora vascular da Serra da Juréia, Municipality of Iguape, São Paulo, Brazil. **Boletim do Instituto de Botânica** v.15, p. 63-124. 2001.

MARCOS FILHO, J.; CÍCERO, S. M.; SILVA, W. R. **Avaliação da qualidade das sementes.** Piracicaba: FEALQ, 1987. 230 p.

MARCOS-FILHO, J. **Fisiologia de sementes de plantas cultivadas**. Piracicaba: Fealq, 2005, 495 p.

MARTIN, C. E. Physiological Ecology of the Bromeliaceae. The Botanical Review v. 60, p.1-82. 1994.

MARTINELLI, G. Manejo de populações e comunidades vegetais: um estudo de caso na conservação de Bromeliaceae. In: ROCHA, F. D.; BERGALLO, H. G.;

SLUYS, M. V. and ALVES, M. A. S. (eds). **Conservation Biology: Essences.** Ed. Rima, São Paulo. 2006. p. 479-503.

MARTINELLI, G.; VIEIRA, C. M.; GONZALEZ, M. P. L.; PIRATININGA, A.; COSTA, A. F. da; FORZZA, R. C. Bromeliaceae from the Brazilian Atlantic Forest: species list, distribution and conservation. **Rodriguésia** v. 59, n. 1, p. 209-258. 2008.

MARTINS, M. S.; RÓZ, A. L.; MACHADO, G. O.. **Atlantic Forest** . 2006. Available at: <http//www.educar.sc.usp.br>. Accessed on: 07/Jul. 2008.

MAYER, A. M.; POLJAKOFF-MAYBER, A. **The germination of the seeds**. v. 3, Oxford: Pergamon Press, 1963. 263 p.

MELO, T. B. As bromélias no paisagismo. **Bromélia** v. 3, p. 3-7, 1996.

MELO, F. P. L.; AGUIAR NETO, A. V.; SIMABUKURO, E. A.; TABARELLI, M. Seedling recruitment and establishment. In: FE RREIRA, A. G. and BORGHETTI, F.

(eds.). **Germination: do básico ao aplicado**. Porto Alegre, Artmed, p. 237-249, 2004.

MELO, M. F. F. and VARELA, V. P. Morphological aspects of fruits, seeds, germination and seedlings of two Amazonian forest species *Dinizia excelsa* Ducke (Angelim Pedra) and *Cedrelinga catenaeformis* Ducke (Cedrorana) Leguminosae: Mimosoideae. **Revista Brasileira de Sementes** n. 28, p. 54-62, 2006.

MENDONÇA, P. G. **Estimation of leaf area of Tillandsia spp. (bromeliacese) and similarity among species based on leaf dimensions**. Dissertation

(Master in Genetics and Plant Breeding) - Universidade Estadual Paulista, Jaboticabal, 2002. 86p.

MORAES, P. L. R. e PAOLI, A. A. S. Morphology and seedling establishment of *Cryptocarya moschata* Nees, *Ocotea catharinensis* Mez and *Endlicheria paniculata* (Spreng.) MacBride - Lauraceae. **Revista Brasileira de Botânica.** v. 22, n. 2, p. 287-295, 1999.

MOREIRA, B. A; WANDERLEI, M. G. L and BAROS, M. A. V. C. **Bromeliads:** ecological importance and diversity. Taxonomy and Mo rphology - Training Course for Monitors. São Paulo: Institute of Botany, 2006. 12p.

NAHOUM, P. Bromeliad. **Revista da Sociedade Brasileira de bromélias** - SBBr. v. 1. p. 1-40 1994.

NOGUEIRA, L. C. **Atlantic Forest:** Origin of the biome, characterization and areas of occurrence . 2008. Available at < http://www.webartigos.com> Accessed on 11/Sep. 2008.

NOVEMBRE, A. D. L. C. **Study of the methodology for conducting the germination test on de-inked cotton seeds (*Gossypium hirsutum* L.)**

mechanically. Thesis (Doctorate in Agronomy) - Luiz de Queiroz College of Agriculture, Piracicaba, SP. 1994. 133 p.

NUNES, J. V. C.: FORZZA, R. C. **Bromelia.** Annals I National Resource Seminar OLIVEIRA, D. M. T. Comparative morphology of seedlings and young plants of native tree legumes: species of *Phaseoleae*, *Sophoreae*, *Swartzieae* and *Tephrosieae*. **Revista Brasileira de Botânica,** n. 24, p. 85-97, 1999.

OLIVEIRA, R. R. Importância das bromélias epífitas na ciclagem de nutrientes da Floresta Atlântica. **Acta Botanica Brasilica,** v. 18, n. 4, p. 793-799, 2004.

OLIVEIRA, D. M. T. Comparative morphology of seedlings and young plants of native tree legumes: species of *Phaseoleae*, *Sophoreae*, *Swartzieae* and *Tephrosieae*. **Revista Brasileira de Botânica.** n. 24, p.85-97, 2001.

PAULA, C. C. Cultivo de Bromélias. **Informe Agropecuário**, v. 26, n. 227, p. 73-84, 2005.

PAULA, C. C.; SILVA, H. M. P. **Cultivo prático de bromélias.** Viçosa, Universidade Federal de Viçosa. 2004. 116p.

PEREIRA, A. R.; PEREIRA, T. S.; RODRIGUES, A. S.; ANDRADE; A. C. S. Seed morphology and post-seminal development of romeliaceae species. **Acta Botanica. Brasilica**. v. 22, n. 4, Oct./Dec. 2008.

QUEIROZ, L. P.; SENA, T. S. N. and COSTA, M. J. L. S. Flora vascular da Serra da Jibóia, Santa Terezinha Bahia. In: O Campo Rupestre. **Sitientibus,** v. 15, p. 27-40, 1996.

REBELO, J. A. **The bromeliads and their surprising survivability .** Available at < http://www.jardimdeflores.com.br> accessed 26/Oct 2004.

ROCHA, P. K. **Desenvolvimento de Bromélias em ambientes protegidos com diferentes alturas e níveis de sombreamento**. Dissertation (Master in Agronomy). ESALQ-USP-Piracicaba, 2002. 90 p.

ROSA, L. S.; FELIPPI, M.; NOGUEIRA, A. C.; GROSSI, F.. Evaluation of germination under different osmotic potentials and morphological characterization of seed and seedling of *Ateleia glazioviana* Baill (Timbó). **Cerne**. n. 11, p 306-314. 2005.

SALVADOR, E. D. Caracterizacao **física e formularizacao de alguns substr atos para o cultivo de alguns plantas ornamentais**. Tese (Doutorado em produção vegetal). Escola Superior de "Agricultura Luiz de Queiroz", Piracicaba, 2000. 148p.

SCATENA, V. L.; SEGECIN, S. E COAN A. I. Seed morphology and post-seminal development of *Tillandsia* L. (Bromeliaceae) from the "Campos Gerais", Paraná , Southern Brazil. **Brazilian Archives of Biology and Technology** v. 49, p. 945-951. 2006.

SILVA, D. A.; KLINK, C. A. Foliation dynamics and p erfil ation of two C4 and one C3 grasses native to the cerrado. **Revista Brasileira de Botânica**, São Paulo, v. 24, n. 4, p. 441-446, 2001.

SMITH, L. B. and DOWNS, R. J. Pitcairnioidea (Bromeliaceae). **Flora Neotropica** Monograph n. 14, Part . 1. New York: OFN-Halfner Press, 1974. 658 p.

SOS MATA ATLÂNTICA. **SOS Projects - Sustainability**. 2006. Available at: <www.ambiente.sp.gov.br/ppma/mataatl1.htm>, 2006. Accessed on: 01 Jul. 2008; <www.apremavi.com.br>, 2006. Accessed on: 01 Jul. 2008; <www.ambientebrasil.com.br>, 2006. Accessed on: 09 Oct. 2008.

STRINGHETA, A. C. O.; SILVA, D. J. H.; CARDOSO, A. A.; FONTES, L. E. F.; BARBOSA, J. G. Germination of seeds and survival of *Tillandsia geminiflora* Brongn seedlings on different substrates. **Acta Scientiarum**. Agronomy Maringá, v. 27, n. 1, p. 165-170, 2005.

TAIZ, L. ; ZEIGER, E. **Plant Physiology**. 2.ed. Sunderland: Sinauer Associates, USA. 2004. 792 p.

TARDIVO, R. C. **Taxonomic revision of Tillandsia L. subgenus Anoplophytum (Beer) Baker (Bromeliaceae),** USP. São Paulo, 2002, 237p.

VAN DER PILJ, L. **Principals of dispersal in higher plants.** 3. ed. Berlin, Springer Verlag. 1982. 215p.

VERDONCK, O.; GABRIELS, R. Substrate requirements for plants. **Acta Horticulture**, v. 221, p. 19-23, 1988.

WALLER, P. L.; HARRINSON, A. M. Estimation of pore space and the calculation of air volume in horticultural substrates. **Acta Horticulture**, n. 294, p. 29-39, 1991.

WENDT, T. **Hybridization and reproductive isolation in *Pitcairnia* (Bromeliaceae)**. Dissertation (Master in Ecology). Federal University of Rio de Janeiro. 1999. 141p.

Chapter 1

FRUIT, SEED AND POST-SEMINAL DEVELOPMENT MORPHOLOGY OF Alcantarea nahoumii (Leme) J.R. Grant. (Bromeliaceae) [1]

[1] Submitted to the Revista Brasileira de Botânica Editorial Committee

FRUIT, SEED AND POST SEMINAL DEVELOPMENT MORPHOLOGY OF *Alcantarea nahoumii* (Leme) J.R. Grant. (Bromeliaceae)

Abstract: *Alcantarea nahoumii* (Helm) J. R. Grant. is rupicolous, considered as a species of high ornamental potential for landscaping, is poorly known, requiring studies for its exploration and conservation. The morphological characterization of *Alcantarea nahoumii* fruits, seeds and seedlings was aimed at its conservation and to increase the understanding of its reproductive strategies. Ripe, dry fruits were collected at the beginning of dehiscence in a natural population in the Serra da Jibóia, municipality of Santa Terezinha, BA, Brazil. Seeds were manually extracted from the fruits, and stored in paper bags, under room temperature of 25 ± 2°C and relative humidity of approximately 50% for a period of three months. For the biometric description of the seeds, length, width and thickness, 100 seeds from approximately twenty-five different individuals, chosen at random, were measured individually with a digital caliper. The weight of one thousand seeds was determined using precision analytical balance. Four samples of 100 seeds were placed to germinate in Plantmax®, in BOD at 25 ºC, with 16 hours of light. The seeds are filiform, measuring 9.6 x 0.9 x 0.4 mm in length, width and thickness, respectively, and a thousand-seed weight of 1.97 g. They display plumose appendages at both ends. The tegument is thin, dark brown to reddish. Germination is cryptoepigeial, beginning on the tenth day after sowing, protruding the cotyledonary sheath with a distal copula. The cotyledon remained encrypted, the neck is indistinct, the primary root de senvolved shortly after 45 days. The first leaf (eophilus) appeared at 16 days, from the lateral expansion of the cotyledonary sheath. The second, third, and fourth leaves appeared at 20, 22, and 45 days, respectively. At this stage, the seedling can be considered normal, with lanceolate leaves and the primary root covered by numerous absorbent hairs. Germination is classified as intermediate, but the slow development of the seedlings limits their attachment to the substrate, which may reduce recruitment in their natural environment and the formation of normal seedlings under cultivation conditions.

Keywords: bromeliad; germination; endemic; propagation.

INTRODUCTION

Seed morphology and post-seed development contributes to the differentiation of taxonomic groups (MORAES and PAOLI, 1999; ROSA et al., 2005), besides helping seed germination and co nservation analyses (ANDRADE et al, 2003) and for studies on regeneration in natural ecosystems (OLIVEIRA, 2001; MELO and VARELA, 2006), since germination or emergence and establishment of seedlings are critical stages in the life cycle of plants (MELO et al., 2004).

The knowledge of the morphological structures of the fruit, seed and seedlings is important for several purposes such as seed analysis laboratories, species identification and differentiation, plant recognition in the field, taxonomy and agriculture (AMORIM, 1996). The fruit and the seed can provide indications about the type of storage, viability and sowing methods (KUNIYOSHI, 1983). In the same way, they contribute to the correct interpretation of germination tests and the realization of scientific works (ARAÚJO and MATOS, 1991), also helping to understand the dynamics of plant populations (OLIVEIRA, 1993; DO NADIO and DEMATTÊ, 2000).

Several studies on fruit and seed morphology have been developed, many of them in a partial way and essentially limited to the description of their most general forms (AMORIM, 1996), mainly concerning Bromeliaceae. However, some authors have developed studies with a great diversity of plant species that provide broad and complex information about these species, which may have commercial applications, being also of fundamental importance for ornithologists, who seek information about food sources used by birds (GROTH and LIBERAL, 1988).

In the Bromeliaceae the cotyledon remains internally in the seed absorbing reserves from the endosperm, keeping the seed attached to the seedling for a certain time (SMITH and DOWNS, 1974). According to Garwood (1996), the haustorial cotyledon is responsible for the absorption and transfer of reserves from the endosperm to seedling growth and is classified in two ways: if the cotyledons eventually emerge and become photosynthesizing, they are considered foliaceous; if the cotyledons are specialized in haustorial organs and do not emerge from the seed, they are classified as reserve cotyledons.

Tillich (2007) in his extensive review of germination terminology of representatives of Poales (Monocotiledoneae), states that the term cryptocotyledonary applies strongly to these plants, because the cotyledon is part foliaceous, emerging from the seminal remains through the release of the cotyledonary sheath; part haustorial, remaining within the teguments and maintaining contact with the reserve tissue.

According to Oliveira (1988) a normal seedling in the subfamily Bromelioideae presents all the necessary structures to continue the adequate development in favorable conditions. The healthy development of the roots, the maximum growth of the first leaf and the emergence of the second leaf are criteria for the identification of normal seedlings in bromeliads, while the expansion of the second leaf indicates the end of the post-seminal development, starting the young plant phase (PEREIRA, 1988).

The Brazilian literature presents a scarcity of specific publications on seed morphology (GROTH and LIBERAL, 1988), despite efforts to produce scientific papers. As the seed trade has expanded, incorporating species of native flora, it is necessary to expand the knowledge about these species.

Alcantarea nahoumii (Leme) J. R. Grant is among a number of other bromeliad species that are listed in the Data Deficient Species of Brazilian Flora (BRASIL, 2008) and vulnerable to extinction (BIODIVERSITY, 2010).

A. nahoumii is a rupicolous species, considered to have high ornamental potential for landscaping. It belongs to the Bromeliaceae family, subfamily Tillandsioideae, native to the Serra da Jibóia, also occurring in other regions in the State of Bahia, Brazil, Area of Tropical Forest al and Campos de Altitude, which approach 800 meters. (MARTINELLI et al, 2008). The Serra da Jibóia has an extension of 6 km and is located in the mun icipality of Santa Terezinha - Bahia, at a latitude of 12°51'S, longitude of 39°2 8'W, altitude of 750-800 m, rainfall should reach at least 1100 mm annually. At the top of the mountain is a gneissic-granitic outcrop, over which a rupestrian field vegetation develops (QUEIROZ et al., 1996).

However, no studies were found that enable item on the conservation and cultivation of *A. nahoumii*, corroborating the classification made by the Ministry of Environment (BRASIL, 2008) that included it in the list of species of Brazilian flora with deficiency of data on the species. The morphology of its seeds and seedlings are necessary knowledge to support taxonomic and ecological studies, in the area of seed technology, in laboratories and nurseries, as well as for studies of natural regeneration.

The objective of the work was to perform the morphological characterization of the fruits, seeds and post-seminal development of *Alcantarea nahoumii,* aiming at its conservation and to broaden the understanding of its reproductive strategies.

MATERIAL AND METHODS

Ripe, dry fruits were collected at the beginning of dehiscence in a natural population in the Serra da Jibóia, municipality of Santa Terezinha, BA.

The fruits were processed and the seeds extracted manually, and then stored in paper bags at an ambient temperature of 25 ± 2°C and relative humidity of approximately 50% for a period of approximately three months. The experiment was conducted in the Seed Analysis Laboratory of the Universidade Federal do Recôncavo da Bahia - UFRB, municipality of Cruz das Almas, Bahia, Brazil.

For the biometric description of the seeds, length, width and thickness, 100 seeds from approximately twenty-five different individuals were randomly chosen and measured individually with a digital pachymeter. The weight of one thousand seeds was determined using an analytical precision scale, using eight repetitions of 500 seeds adapted from Brazil, (1992).

For characterization of the seedlings the seeds were placed to germinate in Gerbox plastic boxes, cleaned and hygienized with ethyl alcohol under not necessarily aseptic conditions, on commercial substrate Plantmax® (200 mL), moistened with 35 mL of water. The boxes were placed in a BOD incubator chamber at a temperature of 25 °C with a sixteen hour period. In each box 100 seeds were placed, in a total of four boxes. Germinated seeds were considered to be when the protrusion of the cotyledonary sheath occurred. A description of the morphology of the seeds and seedlings was made, relating each phenophase to the time and the making of ink drawings on tracing paper.

The illustration and description of seed morphology and organizational structure followed the terminology adopted by Smith and Downs (1974) and Beltrati (1994). The description of the phases of post-seedling development was considered up to the third fully expanded leaf. For the description of the seedlings the terminology of Pereira (1988) was adopted. For illustration and description of fruit morphology and organizational structure we adopted the terminology used by Beltrati (1994) and Barroso et al. (1999). The aspect of the phases of flowering and fruiting, description of the distal portion of the floral scapes were also recorded and illustrated.

RESULTS AND DISCUSSION

The reproductive structure is formed by the floral scape, primary rachis, secondary rachis, rachises, flowers and fruits. The floral scape is completely covered by showy scapular bracts. The inflorescence is formed by the primary rachis, which has internodes, covered by primary bracts, from which secondary rachises, in which the flowers and fruits are formed (Figure 1).

In *A. naouhmii*, the presence of the secondary raceme between the floral bracts was verified (Figure 1). For *Vriesea guttata* this aspect is important in herbarium collections, and the characteristics used in its distinctions are fragile, such as the degree of exposure of the rachis in the inflorescence, position of the floral bract and its size in relation to the septa (SMITH and DOWNS, 1977). These characteristics are dependent on the stage of development of the inflorescence, making identification difficult.

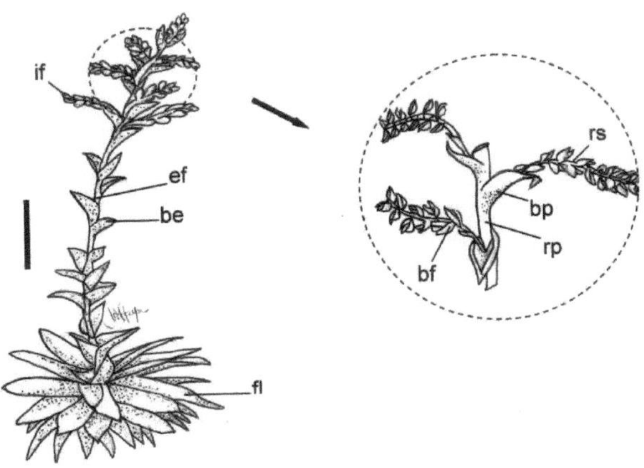

Figure 1. Plant of Alcantarea nahomii (Leme) J. R. Grant ((Bromeliaceae) with inflorescence. if: inflorescence; ef: flower scape al; be: scapular bract; bp: primary bract; bf: floral bract; fl: foliage; rp: primary raceme; rs: secondary raceme. Bar: 50 cm.

The fruits are capsular and trilocular, polyspermic, with septicidal dehiscence, maintaining the floral remains as the sepals and floral bract (Figures 2a, 2b), until the dehiscence phase. T dealing with loculicidal capsular fruits, Barroso et al. (1999) reported that the dehiscence of these fruits occurs along the median vein, passing through the center of the locule and that this median region is perceived by a thickened or protruding line, which represents fragments of septa.

When immature, the fruits present a greenish color, acquiring a dark brown color when the seeds disperse. The exocarp of the fruit had a straw-like appearance. The endocarp was not differentiated from the mesocarp, being perceived as a shiny film on the inner portion of the locules.

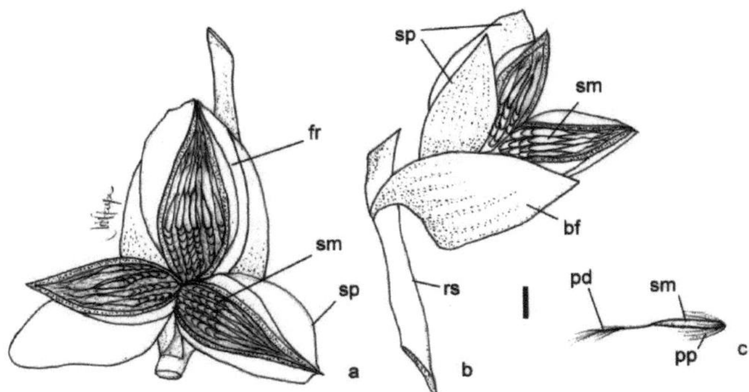

Figure 2. Fruits and seeds of *Alcantarea nahoumii* (Leme) J. R. Grant (Bromeliaceae). a. Frontal view of the septate capsule-type fruit in dehiscence; b. Lateral view of the capsule; c. L ateral view of the seed. bf: floral bract; fr: fruit; pd: distal papus; pp: proximal pa pus; rs: secondary rachis; sm: seed; sp: persistent sepals. Bar: 0,5 cm.

The seeds of A. nahoumii are small (Table 1), light, filiform, (Figure 2c) measuring 9.6 x 0.9 x 0.4 mm in length, width and thickness, respectively, with an average mass of 1,000 seeds of 1.97 g.

Table 1. Seed dimensions of *Alcantarea nahoumii* (Helm) J, R. Grant. (Bromeliaceae).

Variables	Averages (mm) *
Length	9,59±0,83
Width	0,88±0,11
Thickness	0,48±0,07
Weight of one thousand seeds(g)	1,97±0,10

Means followed by standard deviations

The size and shape of the seeds can be very variable, depending on the shape of the ovary, the conditions under which the mother plants grew during seed formation and, of course, the species (MAYER and POLJACOFF- MAYBER, 1963).

Strehl and Beheregaray (2006) point out that in plant systematics, length, width and thickness are tools that can be used to distinguish *Dyckia* species.

The seeds have plumose appendages on both ends, which form papus. The pappus existing in the proximal region of the seed are directed towards the distal region, covering one third of the seed (Figure 2c). Whereas, the papus that arise from the distal region are elongated, corresponding approximately to the same length of the seed (Figure 2c). The tegument is thin, dark brown to reddish, with trichomes from the middle third to half the length of the seed, with longitudinal striations.

The results obtained here corroborate with the morphological characteristics of seeds found in several species of Bromeliaceae. *Pitcairnia encholirioides* L. B. Smith has small, elliptic seeds, approximately 2.0 mm long and 1.0 mm wide (weight of 1,000 seeds 0.162 g), winged, and reddish brown tegument. For *Dyckia pseudococcinea* L. B. Smith the seeds are small, oval-acacia-shaped, discoid, measure approximately 6.0 mm long and 5.0 mm wide (weight of 1,000 seeds 0.345 g), winged and yellowish beige tegument. *Vriesea heterostachys* (Baker) L. B. Smith seeds are small, filiform, measure approximately 6.3 mm long and 1.0 mm wide (weight of 1,000 seeds 0.473 g), winged, with whitish plumose appendages, present only at one end of the seed, brownish tegument (PEREIRA et al., 2008). According to Varadarajan and Gilmartin (1988) and Scatena et al. (2006), the seeds of Bromelioideae have mucilage involving the tegument; while the seeds of Pitcairnioideae and Tillandsioideae have membranous wings and plumose appendages, respectively.

The dispersion is directly related to the different types of fruits present in the family. Winged or plumose seeds, present in capsule-type fruits are aided by the wind and, in the case of succulent berries, whose seeds do not have appendages, dispersal is carried out by animals (MOREIRA et al., 2006). *A. nahoumii*, by presenting plumose seeds at both ends, which form papus, presents characteristics that its dispersion is anemophilic. According to Paula and Silva (2004), the seeds of Pitcairnioideae and Tillandsioideae are small, light and have morphological adaptations that increase the surface/volume ratio, reducing the speed of fall. Generally, they use air currents for transportation in dry periods of the year, facilitating the dispersion of these seeds within the crevices of the rocks, where they find ideal germination conditions. The feathery appendages of epiphytic species help fix the seeds on trunks and tree barks, ensuring the success of their dispersal (VAN DER PIJL , 1982; BENZING, 2000; SCATENA et al., 2006).

Germination of A. nahoumii seeds is crypto-epigeic, beginning on the tenth day after sowing, protruding the cotyledonary sheath with a distal copula. The cotyledonary sheath presents a curving observed by the bending of the margins, and may be more or less fleshy, providing protection to the apical bud (Figures 3a and 3b). The cotyledon remained encrypted, the neck is indistinct, the primary root developed shortly after 45 days (Figures 3a - 3e).

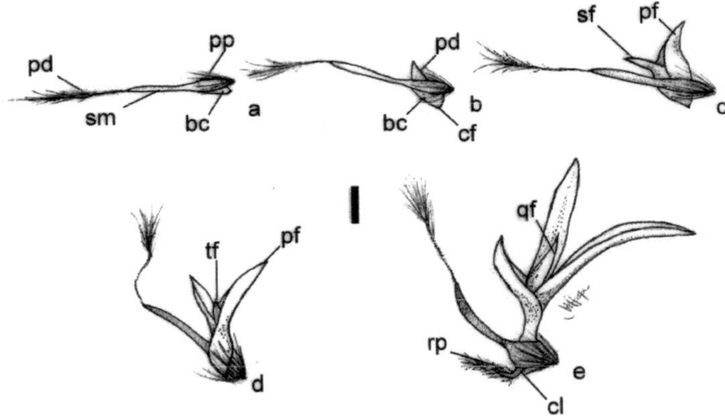

Figure 3. Post-seminal development of *Alcantarea nahoumii* (Leme) J. R. Grant (Bromeliaceae). a. protrusion of cotyledon sheath at 10 days after initiation of germination test; b. Seedling at 16 days; c. Seedling at 20 days; d. Seedling at 22 days; e. Young plant at 47 days: bc: cotyledonary sheath; cf: copula; cl: neck; pd: distal papus; pf: first leaf (eophilus); pp; proximal papus; qf: fourth leaf; rp: primary root; sf: second leaf; sm: seed; tf: third leaf. Bar: 0.5 cm.

The first eophilus appeared at 16 days, from the lateral expansion of the cotyledonary sheath (Figure 3b). The second, third, and fourth eophylls appeared at 20, 22, and 45 days, respectively. At this stage, the seedling can be considered normal, with lanceolate leaves and the primary root covered by numerous absorbent hairs (Figure 3e). Mantovani and Iglesias (2005) verified the appearance of the first leaves in *Aechmea nudicaulis, Neoregelia cruenta* and *Vriesea neoglutinosa* at 8, 11 and 15 days, respectively. It was possible to verify that the Tillandsoideae (*V. neoglutinosa*) also had a slower initial development, as was verified in *A. nahoumii*.

The neck is a well-demarcated structure with a gentle narrowing at the boundary of the hypocotyl and the main root (Figure 3e). As the main root grows and the secondary roots attach themselves to the substrate, the hairs on the neck tend to disappear and, according to Boyd (1932), lose their initial function of attaching the small seedling to the substrate.

Monocots generally present cryptocotyledonary germination, keeping part of the cotyledon inside the seed and part of it emerging from the teguments (DUKE, 1969). The presence of a cotyledonary sheath in the subfamily Bromelioideae, stiffened by vascular bundles that protect the eophilus against friction with the substrate, seems to be related to hypogeous germination, which does not occur with Tillandsioideae and Pitcairnioideae, whose germination is epigeal (BOYD, 1932), as well as in the genera *Aechmea, Quesnelia, Billbergia, Neoregelia, Nidularium, Orthophytum, Canistrum, Portea* and *Streptocalyx* (Bromelioideae) (BENKENDAN and GROB, 1980). Pereira (1988) observed both types of germination for several genera of Bro melioideae.

Pereira et al. (2008) studying six species of Bromeliaceae, observed that these species present cotyledons that rise above the level of the substrate during hypocotyl elongation. Only the species of Tillandsioideae, *Alcantarea imperialis* and *Vriesea heterostachys*, did not show a developed hypocotyl. However, all species were characterized as epigeic germination.

According to Boyd (1932), the characteristics that condition the epigeal germination in Bromeliaceae are related to the tendency to epiphytism and the absence or rare presence of vascular bundles in the cotyledonary sheath. Besides these factors, most species of bromeliads studied require light for seed germination (ME RCIER e GUERREIRO FILHO, 1990; BENZING, 2000).

In *A. nahoumii* it was observed that the first structure to protrude was the cotyledonary sheath exhibiting a distal coppice. Mantovani and Iglesias (2005) found that the protruded structure in germinating *V. neoglutinosa* did not form any type of trichome or absorbent hair. Scatena et al. (2006) observed for *Tillandsia* species that the first structure that emerged during germination was the haustorial cotyledon, without primary root growth.

The seedling type observed for *A. nahoumii* was cryptocotyledonary/epigeal, when the cotyledons remain enclosed in the seminal rsts. Pereira et al. (2008) verified epigeal emergence in *Aechmea blanchetiana* (Baker) L. B. Sm., *Wittrockia gigantea* (Baker) Leme, *Dyckia pseudococcinea* L. B. Sm., *Pitcairnia encholirioides L.* B. Sm, *Alcantarea imperialis* (Carrière) Harms and *Vriesea heterostachys* (Baker) L. B. Sm. whereas Silva (2006) observed crypto-hypogeial emergence in *Bromelia karatas*, indicating that most of the species studied present preferentially epigeial emergence, in agreement with the observations of Pereira (1988).

The pattern of post-seminal development obtained in the present study for this species corroborated with those found by Tillich (2007). According to this author, Bromelioideae has a short cotyledonary sheath, a greatly reduced or non-existent hypocotyl in the seedling, and the primary root grows moderately in length.

The seeds studied, as for the beginning of germination, were classified as intermediate, with an average time of germination between 7 and 14 days, considering that Pereira et al. (2008) in their study with six species from the three subfamilies of Bromeliaceae, classified them as: fast*, Aechmea blanchetiana, Wittrockia gigantea* and *Dyckia pseudococcinea* (mean time < 7 days); intermediate, Alcantarea*. imperialis and* Vriesea *heterostachys (*mean time > 7 < 14 days) and slow, P. encholirioides *(mean time >* 14 days). It should be noted that species classified as fast belong to the Bromelioideae, except *D. pseudococcinea* (Pitcairnioideae); intermediate and slow species belong to the Tillandsioideae and Pitcairnioideae, respectively, Pereira et al. (2008).

Alcantarea nahoumii showed slow post-seedling development and normal seedling formation, indicating limitations in its attachment to the substrate, which represents a reduction in the recruitment of these plants in their natural environment, as well as the formation of young plants under cultivation conditions. In seedling production programs, the knowledge of the time required for germination, development of the seedling and the young plant is of great importance because it allows the planning of the use of space in the beds (PEREIRA et al., 2008).

The results presented in this paper demonstrate that the morphology of fruits, seeds and seedlings of Bromeli aceae are useful tools for taxonomic, ecological and seed technology studies. They can be applied to germination tests, helping laboratories to identify seeds and the duration and description of each phase in the post-seedling development. They can also help in the identification of essential structures and in the diagnosis of normal seedlings, allowing the evaluation of seed lot vigor through seedling vigor.

CONCLUSIONS

1. The fruit dehiscens spontaneously when dry, exposing and releasing the seeds.
2. The small, light seeds make wind dispersal possible and must be placed on the substrate to germinate.
3. Germination is cryptoepigeial, beginning on the tenth day after sowing.
4. The first leaf (eophilus) emerges at 16 days, from the lateral expansion of the cotyledonary sheath.
5. The slow development of the seedlings limits their attachment to the substrate, and may reduce recruitment into their natural environment, as well as the formation of normal seedlings under cultivation conditions.

BIBLIOGRAPHIC REFERENCES

AMORIM, I. L.; **Morphology of fruits, seeds, germination, seedlings and saplings of forest species from the region of Lavras - MG.** Dissertation (Master in Forest Engineering) - Department of Forestry, Federal University of Lavras, MG, 1996. 127p.

ANDRADE, A. C. S.; CUNHA, R.; SOUZA, A. F.; REIS, R. B.; ALMEIDA, K. J. Physiological and morphological aspects of seed viability of a neotropical savannah tree, *Eugenia dysenterica* DC. **Seed Science e Technology**, n. 3, p. 125-137, 2003.

ARAUJO, S. S. MATOS, V. P. Morphology of seeds and seedlings of *Cassia fistula* L. **Revista Árvore** , Viçosa, v.15, n. 13, p. 217-230, 1991.

BARROSO, G. M.; MORIM, M. P.; PEIXOTO, A. L.; ICHASO, C. L. F. **Frutos e sementes:** morfologia aplicada à sistemática de dicotiledôneas . Viçosa: UFV, 1999. 443p.

BEKENDAM, J.; GROB, R. **Manual for evaluation of seedlings in degermination analysis.** Madrid: Instituto Nacional de Semillas y Plantas de Vivero, 1980. 130p.

BELTRATI, C. M. **Morfologia e anatomia de sementes.** Rio Claro: Departamento de Botânica da UNESP, 1994. 108p.

BENZING, D. H. **Bromeliaceae: profile of an adaptive radiation**. Cambridge University Press. 2000. 669p.

BIODIVERSITY. **Consultation on the Revision of the List of Brazilian Flora Threatened with Extinction.** Available at <http://www.biodiversitas.org.br/florabr/grupo3fim.asp>. Accessed on 09/May 2010.

BOYD, L. **Monocotylous seedlings. Morphological studies in the post-seminal development of the embryo.** Transactions and Proceedings of the Botanical Society of Edinburgh, v. 31, p. 5-224. 1932.

BRAZIL, Ministry of Agriculture and Agrarian Reform. **Regras para análise de sementes**. Brasília: SNDA/DNDV/CLAV, 1992. 365p.

BRAZIL. Ministry of Environment. **Official List of Species of Brazilian Flora Threatened with Extinction.** Normative Instruction Nº 6, of September 23, 2008. 55p.

DONADIO, N. M. M.; DEMATTÊ, M. E. S. P. Morphology of fruits, seeds, and seedlings of canafistula (*Peltophorum dubium (*Spreng.) Taub.) and jacarandá-da-Bahia (*Dalbergia nigra* (Vell.) Fr. All. ex Benth.) - Fabaceae. **Revista Brasileira de Sementes**, Brasília, v. 22, n. 1, p. 64-73, 2000.

DUKE, J. A. On tropical tree seedlings 1. Seeds, seedlings, systems and systematics. **Annals of the Missouri Botanical Garden,** v. 56, p. 125-161. 1969.

GARWOOD, N. C. Functional morphology of tropical tree seedlings.. In: SWAINE, M. D. (ed.). **The ecology of tropical forest tree seedlings**. Paris, Man and the Biosphere series, p. 59-129, 1996.

GROTH, D.; LIBERAL, O. H. T. **Catálogo de identificação de sementes**. Campinas: Cargill Foundation n. 1, 1988. 182p.

KUNIYOSHI, Y. S. **Morfologia da semente e da germinação de 25 espécie s arbóreas de uma floresta com araucária**. Dissertation (Master in Forest Sciences) - Department of Forest Engineering, Federal University of Paraná, Curitiba, 1983. 233f.

MANTOVANI, A.; IGLESIAS, R. R. When does the first scale appear? A comparative study on the emergence of absorbing scales in three species of restinga terrestrial bromeliads. **Rodriguesia**, v. 56, n. 87, p. 73-84, 2005.

MARTINELLI, G.; VIEIRA, C. M.; GONZALEZ, M. P. L.; PIRATININGA, A.; COSTA, A. F. da; FORZZA, R. C. Bromeliaceae of the Brazilian Atlantic Forest: species list, distribution and conservation. **Rodriguésia** v. 59, n. 1, p. 209-258. 2008.

MAYER, A. M.; POLJAKOFF-MAYBER, A. **The germination of the seeds**. v. 3, Oxford: Pergamon Press, 1963. 263 p.

MELO, F. P. L.; AGUIAR NETO, A. V.; SIMABUKURO, E. A.; TABARELLI, M. Seedling recruitment and establishment. In: A. G. Ferreira,; F. Borghetti (eds.). **Germination** : do básico ao aplicado. Porto Alegre: Artmed, p. 2 37-249, 2004.

MELO, M. F. F.; VARELA, V. P. Morphological aspects of fruits, seeds, germination and seedlings of two Amazonian forest species *Dinizia excelsa* Ducke (Angelim Pedra) and *Cedrelinga catenaeformis* Ducke (Cedrorana) Leguminosae: Mimosoideae. **Revista Brasileira de Sementes** n. 28, p. 54-62, 2006.

MERCIER, H.; GUERREIRO FILHO, O. Propagacao sexuada de algumas bromélias nativas da Mata Atlântica: efeito da luz e da temperatura na germinação. **Hoehnea,** v.17, p. 19-26. 1990.

MORAES, P. L. R. ; PAOLI, A. A. S. Morphology and seedling establishment of *Cryptocarya moschata* Nees, *Ocotea catharinensis* Mez and *Endlicheria paniculata* (Spreng.) MacBride - Lauraceae. **Revista Brasileira de Botânica** n. 22, p. 287-295. 1999.

MOREIRA, B. A; WANDERLEI, M. G. L.; BAROS, M. A. V. C. **Bromeliads:** ecological importance and diversity. Taxonomy and Mo rphology - Training Course for Monitors and Educators. São Paulo: I nstituto de Botânica, 2006. 12p.

OLIVEIRA, E. C. Morfologia de plântulas. In: PIÑA-RODRIGUES, F. C. M. (Coord.). **Manual de análises de sementes florestais.** Campinas: Gargill Foundation, p. 15-24, 1988.

OLIVEIRA, E. C. Morfologia de plântulas florestais. In: AGUIAR, I.B.; PIÑA-RODRIGUES, F. C. M. (Ed.) **Sementes florestais tropicais**. Brasília: ABRATES, p. 75-214, 1993.

OLIVEIRA, D. M. T. Comparative morphology of seedlings and young plants of native tree legumes: species of *Phaseoleae*, *Sophoreae*, *Swartzieae* and *Tephrosieae*. **Revista Brasileira de Botânica.** n. 24, p.85-97, 2001.

PAULA, C. C.; SILVA, H. M. P. **Cultivo prático de bromélias**. 3 ed. Viçosa: Universidade Federal de Viçosa, 2004. 106 p.

QUEIROZ, L. P.; SENA, T. S. N. and COSTA, M. J. L. S. Flora vascular da Serra da Jibóia, Santa Terezinha Bahia. In: O Campo Rupestre. **Sitientibus,** v. 15, p. 27-40, 1996.

PEREIRA, T. S. Bromelioideae (Bromeliaceae): morphology of post-seminal development of some species. **Arquivos do jardim Botânico do Rio de Janeiro**, Rio de Janeiro v. 29, p. 115-154, 1988.

PEREIRA, A. R.; PEREIRA, T. S.; RODRIGUES, A. S.; ANDRADE, A. C. S de. Seed morphology and post-seminal development of Bromeliaceae species. **Acta Botanica Brasilica**, v. 22, n. 4, p. 1150-1162. 2008.

ROSA, L. S.; FELIPPI, M.; NOGUEIRA, A. C.; GROSSI, F. Avaliação da germinação sob diferentes potenciais osmóticas e caracterização morfológica da semente e plântula de *Ateleia glazioviana* Baill (Timbó). **Cerne**, n. 11, p. 306-314, 2005.

SCATENA, V. L.; SEGECIN, S.; COAN, A. I. Seed morphology and post-seminal development of *Tillandsia* L. (Bromeliaceae) from the "Campos Gerais", Paraná, Southern Brazil. **Brazilian Archives of Biology and Technology,** v. 49, p. 945-951, 2006.

SILVA, E. E. **Northeast native fruits: physiological quality, morphology and cytogenetics.** Dissertation (Master in Agronomy). Federal University of Paraíba. Paraíba, 2006. 110p.

SMITH, L. B.; DOWNS, R. J. *Pitcairnioideae (Bromeliaceae).* **Flora Neotropica**, Monograph Hafner Press, v. 14, n. 1, p. 1-658, 1974.

SMITH, L. B.; DOWNS, R. J. *Tillandsioideae (Bromeliaceae).* **Flora Neotropica**, Monograph Hafner Press, v. 14, n. 2, p. 663-1492, 1977.

STREHL, T.; BEHEREGARAY, R. C. P. Seed morphology of the genus *Dyckia*, subfamily Pitcairnioideae (Bromeliaceae). **Botanical Research** . São Leopoldo, v. 1, n. 57, p. 103-120, 2006.

TILLICH, H. J. Seedling diversity and the homologies of seedling organs in the order Poales (Monocotyledons). **Annals of Botany,** v. 100, p.1413-1429, 2007.

VAN DER PILJ, L. **Principals of dispersal in higher plants.** 3. ed. Berlin, Springer Verlag. 1982. 215p.

VARADARAJAN, G. S.; GILMARTIN, A. J. Taxonomic realignments within the subfamily Pitcairnioideae (Bromeliaceae). **Systematic Botanic,** v. 13, p. 294-299, 1988.

Chapter 2

SEED GERMINATION AND SEEDLING VIGOR OF *Alcantarea nahoumii* (Helm) J.R. Grant. (Bromeliaceae) ON DIFFERENT SUBSTRATES [2]

[2] Submitted to the Editorial Board of the Brazilian Journal of Plant Physiology

SEED GERMINATION AND VIGOR OF PLANTULAS OF *Alcantarea nahoumii* (Leme) J.R. Grant. (**Bromeliaceae**) IN DIFFERENT SOILS.

Abstract: *Alcantarea nahoumii* (Leme) J. R. Grant. is a rupicolous species, which shows high ornamental potential. However, no studies on the germination of its seeds were found. To expand the information about *Alcantarea nahoumii, the* objective was to evaluate the germination behavior and vigor of seeds and seedlings in different substrates, and to adjust the protocol for the analysis of first and final germination counts, aiming its conservation and commercial exploitation. The germination of 4 repetitions of 100 seeds at 25 ºC was done in sand, paper, Plantmax® and vermiculite. Seeds that protruded the cotyledonary sheath were considered germinated and evaluated every two days. The vigor of the seedlings was evaluated at the end of the experiments by measuring the length of the aerial part and roots and the total fresh mass. The density, total porosity and water holding capacity of the substrates were determined. The variables of the germ-nation test were correlated to the attributes of the substrates. Statistical analysis was by Tukey's test, comparing means at 5% probability. Germination occurred at 10 days after sowing, in sand and Plantmax® substrates, and at 14 and 18 days, respectively, in paper and vermiculite. Germination and vigor of seedlings were higher in Plantmax®. The first germination count should be performed at 18 days and the final count at 54 days after sowing.

Keywords: bromeliad; growth; seeds, physical attributes

INTRODUCTION

Alcantarea nahoumii is a rupicolous species, considered to have a high ornamental potential for landscaping. It belongs to the Bromeliaceae family, subfamily Tillandsioideae, native to the Serra da Jibóia, also occurring in other regions in Bahia State, Brazil, Tropical Tropical Forest and Campos de Altitude, which approaches 800 meters. (MARTINELLI et al, 2008).

Sexual propagation is the most efficient commercial method of bromeliad production because of its high natural seed production (ANDRADE e DEMATÊ, 1999). However, the success of this procedure depends on the choice of

The substrate should represent conditions similar to those found in the natural environment of the species and seed vigor. The adequate conditions for germination and/or emergence and development of the root system depend on the absence of pathogens and weeds, adequate pH, texture and structure of the substrate (SILVA et al., 2001).

Seed vigor is the result of a set of characteristics that determine the potential for rapid emergence and uni forme of normal seedlings under a wide range of environmental conditions (AOSA, 1983), and is influenced by edapho-climatic factors, the genetic potential of the species, handling, storage conditions, the size and density of the seed and its age and longevity (LABOURIAL, 1983; BRYANT, 1989; CARVALHO, 1994; CARVALHO and NAKAGAWA, 2000). Vigor is the characteristic that best expresses seed performance (KRZYZANOWSKI and FRANÇA NETO, 2001), and may reflect its potential in relation to crop production (MARCOS FILHO, 2005).

Germination is a process that normally starts with the absorption of water followed by an increase in the metabolic activity of the seed, promoting the intra-seminal growth of the embryo, culminating in the protrusion of the primary root or plumule (LABOURIAU, 1983). The primary root ceases its growth when it reaches about 0.50 cm to 1.00 cm in length (DUARTE, 2007).

Germination is affected by internal factors such as viability and longevity of seeds and also by external factors, such as water availability, oxygen availability, adequate temperature and light (DUARTE, 2007). Under natural conditions, germination regulatory mechanisms can be advantageous to the species, preventing its germination in unfavorable or unstable conditions, thus, survival and germination give the species greater chances of establishing itself in competition with others occupying the same habitat (MAYER and POLJAKOFF- MAYBER , 1963).

The germination evaluation of most cultivated species is usually done in two moments. During the course of the test, the first count is performed, counting the normal seedlings, besides the infected seeds. The first count, obtained together with the germination test, is performed with the purpose of verifying the initial development of the seedlings, and the number of days for this evaluation is variable. As the first count obtains the percentage of normal seedlings, this test can be used as an indication of the relative vigor of the seeds and the lots evaluated (BRASIL, 1992; NAKAGAWA, 1994). The second moment of evaluation is the final count, when normal and abnormal seedlings are evaluated, as well as non-germinated, dormant or dead seeds (BRASIL, 1992).

The characteristics of the substrate is an important aspect to be observed in post-seedling development. A substrate is formed by three phases, the solid phase, which is responsible for the mechanical support of the root system; the liquid phase, which is responsible for supplying the plant with water and nutrients; and the gaseous phase, which guarantees the transport of oxygen and carbonic gas between the roots and the atmosphere (LEMAIRE, 1995).

The importance of the substrate lies mainly in the control of environmental conditions (water, temperature, salinity, nutrients) in the microsite where the seeds are deposited for germination (KÄMPF, 2000). Wendling and Gatto (2002) and Floriano (2004) state that the substrate is important for germination tests, because the capacity to retain water and oxygen, pathogens, the physical structure, among others, can vary from one substrate to another, interfering with germination. The substrate should remain humid, supplying the water needs of the seed, but without excess, which can reduce the availability of oxygen to the seed, reducing metabolic processes, besides facilitating the proliferation of fungus (POPINIGIS, 1985; FIGLIOLIA et al., 1993; PEREZ et al., 1999).

The physical characteristics of the substrates are the most important because the relations between water and air should not be modified over time (VERDONCK et al., 1983). Its quality is determined by the physical characteristics (VERDONCK et al., 1981) and lies in the ability to provide adequate water and air to the plant roots. In most cases the water is sufficient, but the volume of air is a determining property (VERDONCK et al., 1983).

The total porosity is related to the space occupied by air and the capacity to retain and release water to the plants (WALL ER and HARRINSON, 1991). The ideal value for crops indicated by Riviere (1980) is 75% and by Verdonck and Gabriels (1988) is 85%.

Alcantarea nahoumii (Leme) J. R. Grant. is a rupicolous species with high ornamental potential, and possible use in landscaping, due to its durability, remarkable beauty and exuberance. The roset a has light green leaves with a diameter between 0.4 and 1.2 m and a height of about 1.0 m (GRANT, 1995). The scapular bracts are red-orange and the floral bracts light yellow, making it an element of impact in the garden, whether used alone or in groups and in conjunction with other species of bromeliads.

There is a lack of information on seed germination, substrate type, growth, requirements and tolerances of native plants, as already observed by Labouriau (1983). A number of professionals have dedicated themselves to determining standards to be adopted for the detection of the quality of seed lots of native species, with a view to their storage and utilization, and their conservation in germplasm banks (CARMONA, 1999).

In order to contribute to the expansion of information about *Alcantarea nahoumii, the* objective was to evaluate the germination behavior and vigor of seeds and seedlings on different substrates, in the presence of light and in the dark, and to determine the time of the first and final germination counts, with a view to their conservation and commercial exploitation.

MATERIAL AND METHODS

The experiment was conducted in the Seed Analysis Laboratory of the Universidade Federal do Recôncavo da Bahia (UFRB), municipality of Cruz das Almas, Bahia, Brazil, using seeds of *Alcantarea nahoumii (*Leme) J. R. Grant. (Bromeliaceae), collected from fruits at the beginning of dehiscence, in a natural population in the Serra da Jibóia, municipality of Santa Terezinha, BA. Area of Tropical Forest and Campos de Altitude, approaching 800 meters. (MARTINELLI et al, 2008) at a latitude of 12°51'S, longitude 39°28'W, altitude 750-800 m, rainfall should reach at least 1,100 mm annually. (QUEIROZ et al., 1996).

The seeds were extracted manually, and then stored in paper bags under room temperature of 25 ± 2˚C and relative humidity of approximately 50% for a period of approximately three months.

For the physical characterization of the seeds, 100 seeds from approximately twenty-five different individuals were randomly chosen and measured individually with a digital pachymeter. The weight of one thousand seeds was determined using an analytical precision scale, using eight repetitions of 500 seeds, according to a methodology adapted from Brazil (1992).

The water content of the seeds on a wet basis was determined by the oven method at 105˚C ± 3˚C for 24 hours (Brazil, 1992), using four repetitions of 100 seeds.

The substrates were sand, paper, Plantmax® and vermiculite. Sowing was performed on the substrate in Gerbox-type plastic boxes containing two sheets of filter paper, pre-moistened with a volume of distilled water equivalent to 2.5 times the mass of the paper (Brazil, 1992), washed and sterilized sand, Plantmax® and vermiculite. The last three substrates were moistened with 50.0 mL; 35.0 mL and 42.0 mL of distilled water, respectively, as moisture was perceived at the bottom of the container. The experimental design for the germination test was entirely randomized, with four repetitions and 100 seeds per repetition.

The boxes containing the seeds were kept in a BOD-type germination chamber at a temperature of 25 °C, photoperiod of 16 hours. The germination was evaluated every two days, until the process stabilized, and the results were expressed in percentage, and the germination speed index (IVG) was calculated (MAGUIRE, 1962) using the formula:

IVG = $\sum\left(\frac{gi}{di}\right)$

In which:
gi: number of seeds germinated at the ith count;
di: number of days to germinate from the hard seed to the ith count.

Seeds that presented protrusion of the cotyledon sheath were considered germinated (Pereira, 1988). We also evaluated the percentage of firm and dead seeds (Brazil, 1992); the average length (mm) of the aerial part, determined with a digital pachymeter from the neck to the end of the largest leaf; the average length (mm) of the roots, determined with a digital pachymeter from the neck to the end of the main root; the total seedling fresh mass (mg), determined by weighing on precision analytical scales. After the end of the germination evaluations the first count was defined at 18 days and the final count at 54 days, expressed in percentage. For the analyses of the average length of the aerial part, average root length and the total fresh mass of the seedling, 25 seedlings/treatment were used, with each treatment consisting of four repetitions.

The moment of the first count was determined when about 50% of maximum germination was observed for the best treatments. The average period of these treatments was indicated. The final count was determined when germination of most treatments reached stability.

Analyses of the physical attributes of the substrates such as density, total porosity and water-holding capacity were performed using the methodology described by Fretz et al. (1979).

Table1 . Physical attributes* of substrates used for germination of seeds of *Alcantarea nahoumii (*Leme) J, R. Grant. (Bromeliaceae).

Substrates	Density (g.cm-3)	Total Porosity (%)	Holding capacity of water (%)
Sand	1,80	20,44	12,68
Paper	0,32	80,00	57,50
Plantmax®	0,65	69,62	26,27
Vermiculite	0,24	82,30	63,93

Statistical analysis of germination data and seedling vigor was performed using Tukey's test, comparing means at 5% probability. The statistical program SISVAR (Ferreira 2000) was used to perform the analyses. A simple Pearson correlation was performed between the variables of germination and the attributes of the substrates

RESULTS AND DISCUSSION

The water content of the seeds was approximately 13% (Table 2), which was adequate for germination, however, a little above what would be considered ideal for conservation of orthodox seeds, because when they are dehydrated their metabolism is reduced to minimum levels, allowing them to survive to environmental stresses (CASTRO et al., 2004). Marcos Filho (2005) reports that water contents between 10% and 12% allow the maintenance of germination for a period of six to eight months for most species.

The thousand seed weight of *A. nahoumii* was 1.97 g (Table 2). Silva (2006) found that *Bromelia karatas had a thousand seed weight of* 47.80 g. For *Nidularium inncentii* and *N. procerum* the thousand seeds weight was 2.14 g and 1.99 g respectively (PEREIRA, 2009). The weight of one thousand seeds, which is generally used to calculate the sowing density and the weight of the work sample, is information that gives an idea of the quality of the seeds, as well as their state of maturity and sanity (BRASIL, 1992).

Table 2. Physical characteristics of the seeds of *Alcantarea nahoumii* (Leme) J, R. Grant. (Bromeliaceae).

Variables	Averages *
Water content (%)	12,96±0,41
Weight of one thousand seeds(g)	1,97±0,10

Means followed by standard deviations

A. nahoumii started germination at 10 days after sowing, in sand and Plantmax® substrates, which remained the same during almost all the time evaluated. Only at 60 days after sowing there was a significant difference between the means of germination. In the substrates paper and vermiculite, germination started at 14 and 18 days, respectively (Figure 1).

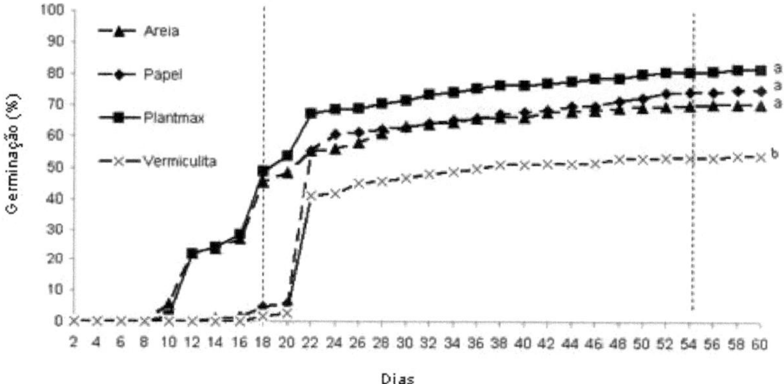

Figure 1. Germination behavior of *Alcantarea nahoumii* (Leme) J. R. Grant (Bromeliaceae) seeds on different substrates, over of 60 days. Means followed by different letters, in the last evaluation, differ from each other at 5% probability by the Tukey test.

The first count was determined at 18 days (Figure 1), since even in the new Rules for Seed Analysis (Br asil, 2009) there are no recommendations for the species studied. The first count was a determination performed together with the germination test serving to evaluate seed vigor (VIEIRA and CARVALHO, 1994). The final count was determined at 54 days, when the germination process stabilized and can be used to express the total percentage of germinated seeds.

A significant difference was observed among substrates for first count, in which the highest averages occurred in the substrates Plantmax® and Sand respectively. Except for vermiculite, the other substrates did not differ statistically among themselves regarding the germination percentage. No firm seeds were observed in Plantmax®, (Table 3), which implies that all seeds that did not die germinated (81.75%).

The first count test makes it possible to infer on the physiological quality of the seeds. Duarte, (2007) working with Dyckia goehringii, observed that the first count of the germination test distinguished large seed lots from small seed lots, regarding their physiological quality.

Table 3. Germination percentage at first count (18 days), germination speed index (GVI), firm and dead seeds of *Alcantarea nahoumii* (Leme) J. R. Grant (Bromeliaceae), at 60 days after germination.

Treatment	First Count(%)	Germination (%)	IVG*	Seeds*	
				Firm (%)	Dead(%)
Sand	45,5 b	70,7 a	43,9 a	14,5 b	14,7 a
Paper	4,8 c	75,2 a	34,8 b	19.5 ab	5,2 a
Plantmax®	49,0 a	81,75 a	49,4 a	0,0 c	18,2 a
Vermiculite	1,8 d	54,0b	25,3 c	30,0 a	16,0 a

Means followed by different letters in the columns differ at 5% probability by Tukey's test.

The GVI for Plantmax® and Sand differed statistically between paper and vermiculite (Table 3). The speed of germination is one of the oldest tests of seed vigor and can discriminate seed lots that show similar germination, being considered more vigorous those lots that show faster germination (NAKAGAWA, 1994). According to Silva (2006), native species, in general, show slow growth, hence the importance of defining a substrate that promotes the speed and uniformity of germination, combined with temperature, adequate storage period and seeds of good physiological quality.

The vigor represented by the GVI can determine the establishment of the *D. goheringii* seedling in the field (DUARTE, 2007). Hernandez et al. (1999) reported that *Tillandsia guatemalensis*, showed germination under laboratory conditions above 93%, however, under natural conditions, with the occurrence of torrential rains, many plants were washed away before settling in the substrate, leading to a reduction of more than half of the young plants.

The growth of *A. nahoumii* seedlings was small 60 days after the installation of the experiment, not exceeding 13.0 m in length of the aerial part and 2.6 mm in the roots (Table 4). The use of Sand and Plantmax® presented significant variations for average length of the aerial part, in relation to vermiculite and paper. Root growth in Plantmax® was significantly greater than those germinated in vermiculite (Table 4).

Table 4. Mean aboveground and root length and fresh mass of *Alcantarea nahoumii (*Leme) J. R. Grant (Bromeliaceae) seedlings at 60 days after germination.

Treatment	Aerial Part (mm)	Root (mm)	Mass (mg)
Sand	12,15 a	1,73 ab	10,16 b
Paper	7,56 c	1.51 bc	6,80 b
Plantmax®	12,76 a	2,53 a	14,35 a
Vermiculite	9,91 b	0,76 c	10,07 b

* Means followed by different letters in the columns differ at 5% probability by Tukey's test.

The results verified in Table 4 demonstrate that the substrate affects the partition of assimilates in the initial development phase of *A. nahuomii.* The total fresh mass of the developed seedlings was higher in Plantmax®

The lowest density and the highest mean total porosity and water retention capacity (82.30% and 63.93%) were observed for vermiculite. Riviere (1980) considers ideal for cultivation, substrates that have total porosity of 75%, while Verdonk and Ga briels (1988) consider ideal substrates with 85% of total porosity, however, considering the results of germination and vigor of seedlings of *A. nahoumii*, it appears that this attribute was not favorable for determining the appropriate substrate for evaluation of germination in this species.

There were no significant variations among the tested variables in the correlation analysis (Table 5). Germination and first count correlated from moderate (>0.50) to strongly positive (>0.80) (SANTOS, 2007) with root length, indicating tendencies for higher germination in the first count and in the final count with higher root development. Moderate (> -0.50) to strongly negative (> -0.80) correlation for percent firm seeds with germination at first count, with final germination and with GVI. The aerial part (AP) showed only strongly positive correlation with the first count, indicating that seeds that germinate faster, incorporate more biomass. Seedling fresh mass showed a strong positive correlation with aboveground growth, indicating that this region contributes more to the accumulation of biomass than the roots. The mortality of the seeds showed low correlation with the other variables analyzed.

Table 5. Pearson's simple correlation between the variables of germination and vigor of *Alcantarea nahoumii* (Leme) J. R. Grant (Bromeliaceae) seedlings.

Variables*	GERM	P CONT	IVG	FIRM	MORTAS	PA	ROOT	MASS
GERM	1,00	0,86	0,52	-0,77	-0,27	0,25	0,78	0,31
P CONT		1,00	0,81	-0,73	0,36	0,84	0,68	0,66
IVG			1,00	-0,82	0,01	0,62	0,86	0,57
FIRM				1,00	-0,41	-0,62	-0,73	-0,56
MORTAS					1,00	0,58	-0,02	0,39
PA						1,00	0,50	0,82
ROOT							1,00	0,65
MASS								1,00

* Germination: GERM; First count: P CONT; Germination speed index: IVG; Firm seeds: FIRM; Dead seeds: DEAD; Aboveground length: AP; Root length: Root; Fresh mass of plantlets: MASS.

There was no correlation between substrate attributes and the variables tested in the correlation analysis (Table 6). There was a moderate to strong positive correlation (>0.50) between substrate density and the variables tested, except for dead seeds.

Density showed a negative correlation with firm seeds, with this behavior being attributed to a lower availability of water for the seeds and consequently a lower germination percentage

Table 6. Simple Pearson's correlation between substrate attributes and variables of germination and vigor of *Alcantarea nahoumii* (Helm) J. R. Grant (Bromeliaceae) seedlings.

Variables*	GERM	P CONT	IVG	DEAD	FIRMS	PA	ROOT	MASS
DENS	0,57	0,77	0,74	-0,79	0,38	0,73	0,77	0,81
PT	-0,15	-0,64	-0,49	-0,14	-0,14	-0,53	-0,23	0,10
CRA	-0,48	-0,91	-0,81	0,62	-0,25	-0,76	-0,62	-0,45

* Density: DENS; Total porosity: PT; Water holding capacity: CRA; Germination: GERM; First count: P CONT; Germination speed index: IVG; Seeds

roots: Root; Seedling fresh mass: Mass.

The water-holding capacity (WBC) of the substrates showed moderate correlation with germination and strong negative correlation with the first count, with the GVI and with the growth of the aerial part and the roots (Table 6).

For vermiculite, despite the high CRA (Table 1), this is not available for the seeds due to the granulometry not favoring the seed-substrate contact. According to Oliveira Junior and Delistoianov (1996) this reduces the speed of water absorption. Salvador (2000) points out that the water available to plants in a substrate is more important than the CRA, and by the results observed in this study, this attribute of the substrate can also be considered important for germination of *A. nahoumii.*

According to Fernandes et al. (2006) a greater proportion of small particles in the substrate decreases the percentage of seed germination by hindering the absorption of water in the first days after sowing and by impairing aeration for the roots. Carneiro and Guedes (1992) reaffirm that the greater the contact of the seeds with the substrate, the shorter the time required for total germination to be achieved, provided that the essential elements are available and in adequate quantities.

The study contributed to the development of germination analysis, indicating substrate attributes for germination, the moment of the first count and the final count of germination of *A. nahomii,* and may serve as a reference for other species that present similar behavior. These advances will improve laboratory practices, resulting in efficiency in routine work.

CONCLUSIONS

This study contributed to the development of germination analysis, indicating substrate attributes for germination, the moment of the first and final germination counts for *A. nahomii,* and may serve as a reference for other species with similar composition. Such advances will improve laboratory practices, resulting in efficiency in routine work.

BIBLIOGRAPHIC REFERENCES

ANDRADE, F. S. A.; DEMATÊ, M. E. S. P. Estudo sobre produção e comercialização de bromélias nas regiões sul e sude ste do Brasil. **Revista Brasileira de Horticultura Ornamental**, v. 5, n. 2, p. 97-110, 1999.

AOSA, Association of Official Seed Analysts. **Seed vigor testing handbook**. AOSA. 1983. 88p. (Contribution n° 32 To the handboo k on Seed testing).

BRAZIL, Ministry of Agriculture and Agrarian Reform. **Regras para análise de sementes**. Brasília: SNDA/DNDV/CLAV, 1992. 365 p.

BRAZIL, Ministry of Agriculture, Livestock and Food Supply. **Regras para análise de sementes**. Brasília: Mapa/ACS, 2009. 399 p.

BRYANT, J. A. **Physiology of the seed**. Trad. KRAUS, J.; TRENCH, K. U. S., São Paulo. EPU. 1989. 86p.

CARMONA, R.; MARTINS, C. R.; FÁVERO, A. P. Características de sementes de gramíneas nativas do cerrado. **Pesquisa Agropecuária Brasileira**, Brasília, v. 34, n. 6, p.1067-1074, jun. 1999.

CARNEIRO, J.W.P.; GUEDES, T.A. Influence of Stevia (Stevia rebaudiana (Bert.) Bertoni) seeds contacts on the substrate evaluated by the Weibull function. **Revista Brasileira de Sementes**. Brasília, v. 4, n. 1, p. 65-68. 1992.

CARVALHO, N. M. O conceito de vigor em sementes. In: VIEIRA, R. D.; CARVALHO, N. M. **Teste de vigor em sementes**. Jaboticabal: FUNEP, p. 1-30, 1994.

CARVALHO, N. M.; NAKAGAWA, J. **Sementes ciência, tecnologiae produção.** Jaboticabal: Funep, 2000. 588p.

CASTRO, R. D.; BRADFORD, K. J.; HILHORST, H. W. M. Seed development and water content. In: FERREIRA, A. G.; B ORGHETTI, F. (Orgs.). **Germination** : do básico ao aplicado. Porto Alegre: Artmed, p. 5 1-67, 2004.

DUARTE, E. F. **Caracterização, qualidade fisiológica de sementes e crescimento inicial de *Dyckia goerhringii* Gross & Rauh, bromélia nativa do cerrado.** Thesis (Doctorate in Agronomy). Universidade Federal de Goiás. Goiás, 2007. 200p.

FERNANDES, C.; CORÁ, J. E.; BRAZ, L. T. Alterações nas propriedades físicas de substratos para cultivo de tomate cherry, em função de sua reutilização. **Horticultura brasileira**, v. 24, n. 1, p. 94-98, 2006.

PEREIRA, T. S. Bromelioideae (Bromeliaceae): morphology of post-seminal development of some species. **Arquivos do jardim Botânico do Rio de Janeiro**, Rio de Janeiro v. 29, p. 115-154, 1988.

FIGLIOLIA, M. B.; OLIVEIRA, E. C.; PIÑARO DRIGUES, F. C. M. Análise de sementes. In: AGUIAR, I. B.; PIÑA-RODRIGUES, F. C. M.; FIGLIOLIA, M. B. Sementes **florestais tropicais.** Brasília: ABRATES, p. 137-174, 1993.

FLORIANO, E. P. **Germination and dormancy of forest seeds.** Caderno Didático n. 2, 1st ed., Santa Rosa: Anorgs, 2004. 19p.

FRETZ, T. A. READ, P. E. PEELE, M. C. **Plant propagation Lab Manual**. Mineapolis: Burgess Publishiny Company, 1979, 317 p.

GRANT, J. R. Monocot family: Bromeliaceae RANK: species *Alcantarea nahoumii* (LEME) J. R. GRANT. **Tropische und Subtropische Pflanzenwelt.** n. 91, p. 13, 1995.

HERNÁNDEZ, J. C. C., WOLF, J. H. D., GARCÍA-FRANCO, J. G. & GONZÁLEZ-ESPINOSA. The influence of humidity, nutrients and light on the establishment of the epiphytic bromeliad Tillandsia guatemalensis in the highlands of Chiapas, Mexico. **Revista de Biologia Tropical** v. 47, p. 763-773. 1999.

KÄMPF, A. N. **Produção commercial de plantas ornamentais**. Guariba: Agropecuária, 2000. 254 p.

KRZYZANOWSKI, F. C.; FRANÇA-NETO, J. B. Vigor de se mente. **Informativo ABRATES,** Brasília, v. 11, n. 3, p. 81-84, 2001.

LABOURIAL, L. G. **The Germination of Seeds**. Washington: General Secretariat of the Organization of American States, 1983. 173p.

LEMAIRE, F. Physical, chemical and biolological properties of growing medium. **Acta Horticulturae,** Wageningen, n. 396, p. 273-284, Sept. 1995.

MAGUIRE, J. D. Speed of germination-aid in seletion and evaluation seedling emergence and vigor. **Crop Science**, Madison, v. 2, p. 176-177, 1962.

MARCOS-FILHO, J. **Fisiologia de sementes de plantas cultivadas**. Piracicaba: Fealq, 2005, 495 p.

MARTINELLI, G.; VIEIRA, C. M.; GONZALEZ, M. P. L.; PIRATININGA, A.; COSTA, A. F. da; FORZZA, R. C. Bromeliaceae from the Brazilian Atlantic Forest: species list, distribution and conservation. **Rodriguésia** v. 59, n. 1, p. 209-258. 2008.

MAYER, A. M.; POLJAKOFF-MAYBER, A. **The germination of the seeds**. v. 3, Oxford: Pergamon Press, 1963. 263 p.

NAKAGAWA, J. Testes de vigor baseada na avaliação das plântulas. In: VIEIRA, R. D.; CARVALHO, N. M. **Testes de vigor em sementes. Jaboticabal**: FUNEP, 1994. p.49-85.

OLIVEIRA JR, R. S.; DELISTOIANOV, F. Sowing depth and dormancy breaking methods affecting germination and emer gence of Desmodium purpureum (Mill) Fawc. Et rend. (Leguminosae - Papilionoidea). **Revista Brasileira de Botânica** , v.19, n.2, p.221-225, 1996.

PEREIRA, T. S. Bromelioideae (Bromeliaceae): morphology of post-seminal development of some species. **Arquivos do jardim Botânico do Rio de Janeiro**, Rio de Janeiro v. 29, p. 115-154, 1988.

PEREIRA, C. **Fruit harvest point, quality and seed storage of *Nidularium innocentii* (Lem.) and *Nidularium procerum* (Lindm)**. Dissertation (Master of Science), Universidade Fe deral do Paraná, Curitiba, 2009. 78p .

PEREZ, S. C. J. G. A.; FANTI, S. C.; CASALI, C. A. Influence of storage, substrate, early aging and seeding depth on canafistula germination. **Bragantia**, v. 58, n. 1, p. 57-68, 1999.

POPINIGIS, F. **Fisiologia da semente**. Brasília: Agiplan, 1985. 289p.

QUEIROZ, L. P.; SENA, T. S. N. and COSTA, M. J. L. S. Flora vascular da Serra da Jibóia, Santa Terezinha Bahia. In: O Campo Rupestre. **Sitientibus,** v. 15, p. 27-40, 1996.

RIVIERE, L. M. Importance of physical characteristics in the choice of substrates for above-ground crops. **Revue Horticole**, v. 209, pp. 23-27, 1980.

SALVADOR, E. D. Caracterizacao **física e formularizacao de alguns substr atos para o cultivo de alguns plantas ornamentais**. Tese (Doutorado em produção vegetal). Escola Superior de "Agricultura Luiz de Queiroz", Piracicaba, 2000. 148 p.

SANTOS, C. **Estatísitica descritiva - Manual de Auto-aprendizag em**. Lisboa: Edições Sílabo. 2007. Available at <http://estatisticax.blogspot.com/2008/04/ coefficient-de-correlacao-de pearsonr.html>. Accessed 03/jun. 2010.

SILVA, R. P. da; PEIXOTO, J. R.; JUNQUEIRA, N. T. V. Influence of different substrates on the development of seedlings of passion fruit tree (Passiflora edulis Sims f. flavicarpa Deg). **Revista Brasileira de Fruticultura,** Jaboticabal, v. 23, n. 2, p. 377-381, 2001.

SILVA, E. E. **Frutiferas Nativos do Nordeste: qualidade fisiológi ca, morfologia e citogenética.** Dissertation (Master in Agronomy). Federal University of Paraíba. Paraíba, 2006. 110p.

VERDONCK, O.; De VLEESCHAUWER, D.; De BOODT, M. The influence of the substrate plant growth. **Acta Horticulture** , v. 126, p. 251-258, 1981

VERDONCK, O.; PENNINK, A.; D.; De BOODT, M. The physical properties of different horticultural substrats. **Acta Horticulture**, n.150, p. 155-160, 1983.

VERDONCK, O.; GABRIELS, R. Substrate requirements for plants. **Acta Horticulture**, v. 221, p. 19-23, 1988.

VIEIRA, R. D.; CARVALHO, N. M. **Testes de vigor em sementes**. Jaboticabal:

FUNEP, 1994. p.103-132.

WALLER, P. L.; HARRINSON, A. M. Estimation of pore space and the calculation of air volume in horticultural substrates. **Acta Horticulture**, n. 294, p. 29-39, 1991.

WENDLING, I.; GATTO, A. **Substratos, adubação e irrigação na produção de mudas**. Viçosa: Aprenda Fácil, 2002. 145 p. (Coleção jardinagem e paisagismo; Série produção de mudas ornamentais, 2)

Chapter 3

PHYSIOLOGICAL BEHAVIOR OF SEEDS OF Alcantarea nahoumii (Helm) J.R. Grant. (Bromeliaceae) UNDER DIFFERENT TEMPERATURES [3]

[3] Submitted to the Editorial Board of the Brazilian Journal of Plant Physiology

PHYSIOLOGICAL *BEHAVIOR OF SEEDS OF Alcantarea nahoumii* (Leme) J.R. Grant. (Bromeliaceae) UNDER DIFFERENT TEMPERATURES

Abstract: *Alcantarea nahoumii* (Leme) J. R. Grant. is rupicolous, considered as a species of high ornamental potential for landscaping. Studies on the adequate temperature for germination of its seeds are scarce. The objective was to determine the thermal parameters and the time required to perform germination tests. Germination under constant temperatures of 20, 25, 30 and 35 °C were evaluated. Considered as germinated those that protruded the cotyledonary sheath. Germination was evaluated every two days. The experimental design for the germination test was entirely randomized, with four repetitions, being 100 seeds/repetition. The length of the aerial part and root, and the total fresh mass of the seedlings were evaluated, with four repetitions of 25 seedlings/per repetition. The statistical analysis of the data was done by Tukey's test, comparing the means at 5% probability. The seeds of *Alcantarea nahoumii* present restrictions for satisfactory germination under a temperature of 35°C. To obtain more vigorous seedlings of *A. nahoumii*, seeds should be used at 25°C. The germination test can be evaluated at 18 and 54 days after sowing.

Keywords: bromeliad; germination; vigor; first count.

INTRODUCTION

According to Salisbury and Ross (1991), the growth of plant species is adapted to the temperatures of their natural environment, being sensitive to temperature changes within a few degrees. Moreover, different tissues and organs of the same plant can be affected differently by temperature. This temperature influences all stages of plant development, from the germination of seeds, altering the speed and the final percentage of the process, affecting mainly the absorption of water and all the biochemical reactions that characterize germination (POPINIGIS, 1985).

Generally, seeds respond to temperature to germinate, and in research it is necessary to use extreme temperatures to determine the optimum range, i.e. the temperature range at which maximum germination occurs in the shortest time (LAB OURIAL, 1983).

The temperature reduction leads to a decrease in the speed of chemical reactions, making the biomembranes more rigid and requiring a greater amount of energy to activate biological processes. As a consequence, the membranes become solid-gel, causing an increase in permeability and reducing selectivity, besides increasing the activation energy value of the enzymes attached to them (LARCHER, 2006).

Most cultivated species require temperatures between 20 °C and 30 °C for germination of their seeds (MARCOS FILHO, 2005), a fact that also occurs for species of rupestrian fields, such as species of the genus *Xyris* (GOMES and FERNANDES, 2002).

Smith and Downs (1974) considered that knowledge of the temperature required for germination of bromeliad seeds can help in understanding the distribution of the species. Pinheiro and Borghetti (2003), working with *Aechmea nudicaulis* and *Streptocalyx floribundus*, species from restinga, found that their seeds germinated best between 20 °C and 30 °C, in the presence of light.

The extinction of endemic species or not, is related to the loss of their habitats, either by the expansion of agricultural frontiers, by the occurrence of fire or by other factors no less important (D UARTE, 2007). Siqueira Filho and Tabareli (2006) reported the disappearance of populations of about 20 species of Bromeliaceae from an area of Atlantic Forest, due to the destruction of their habitats.

Alcantarea nahoumii is an endemic species of the Atlantic Forest, making the planning and implementation of research aimed at its conservation and cultivation an urgent necessity.

There is an important demand for research in the area of seed technology, since there are few studies on the physiological quality of seeds and tests that can correctly assess the germination capacity. To evaluate the quality of a given lot of seeds in the laboratory, it is necessary to have a germination standard for each species, because each presents seeds with distinct characteristics regarding their physiological and germination behavior (WIELEWICKI et al., 2006). Thus, research that contributes to the generation of technical knowledge of native species, such as *A. nahoumii,* and methods for a standardization of vigor and germination tests of species not worked by research are essential (ABDO and PAULA, 2006).

One of the most widely used tests in seed analysis is germination, which aims to obtain information about seed quality, both for sowing in the field and for use with other information that can compare seed lots (BARROS et al., 2002).

Studies on the adequate temperature for germination of *A. nahoumii* seeds are scarce. Borghetti and Ferreira (2004) point out that such studies are important to provide not only useful information for seed technology but also for understanding the ecophysiology of plant species. Temperature and light are the main environmental factors that affect seed germination in soil, provided that water and oxygen are available. For many plant species, when adequate light and humidity conditions are provided, the prevailing temperature determines not only the fraction of seeds that germinates but also the speed of germination (ANDRADE, 1995).

The study aimed to determine the thermal parameters and the time required to perform seed germination tests.

MATERIAL AND METHODS

The experiment was conducted in the Seed Analysis Laboratory of the Universidade Federal do Recôncavo da Bahia - UFRB, using seeds of *Alcantarea nahoumii* (Leme) J. R. Grant. obtained from dry fruits in the beginning of dehiscence, collected from plants of a population in the Serra da Jibóia, in the municipality of Santa Terezinha, Bahia, Brazil, an area of Tropical Forest and Campos de Altitude, which is close to 800 m. (MARTINELLI et al, 2008) at a latitude of 12°51'S, longitude 39°28'W, altitude 750-800 m, rainfall should reach at least 1,100 mm annually. (QUEIROZ et al., 1996).

The fruits were processed and the seeds were extracted manually, and then stored in paper bags at an ambient temperature of 25 ± 2°C and relative humidity of approximately 50% for a period of approximately fifteen days

The seeds were placed to germinate in plastic Gerbox boxes, cleaned and sanitized with ethyl alcohol under not necessarily aseptic conditions, on commercial substrate Plantmax® (20 mL), moistened with 35 mL of water. The boxes were placed in a BOD-type incubator at a constant temperature of 20 °C, 25 °C, 30 °C and 35 °C, with a 16-hour photoperiod.

The experimental design for the germination test was entirely randomized, with four repetitions and 100 seeds.

The germination evaluation was cumulative and performed every two days until the process stabilized. The data expressed in average percentage of germinated seeds for each treatment. The germination speed index (GVI) was obtained according to Maguire (1962), using the formula:

$$IVG = \sum\left(\frac{gi}{di}\right)$$

In which:
gi: number of seeds germinated at the ith count;

di: number of days to germinate from sowing to the ith count.

The moment of the first count was determined when about 50% of maximum germination was observed for the best treatments. The average period of these treatments was indicated. The final count was determined when germination of most treatments reached stability.

For the analyses of average aerial part length, average root length and total seedling fresh mass, 25 seedlings/treatment were used, with each treatment consisting of four repetitions.

The water content (%) and dry mass (DM) of 100 seeds (g) were obtained by the oven method, using a temperature of 105+3 °C for 24 hours (h) (Brazil, 1992), and analytical balance. The results were expressed as an average percentage on a wet basis.

Statistical analysis of germination data and seedling vigor was performed using Tukey's test, comparing means at 5% probability. The statistical program SISVAR (Ferreira 2000) was used to perform the analyses.

RESULTS AND DISCUSSION

The water content of the seeds was approximately 15% (Table 1), which was adequate for germination. Marcos Filho (2005) reports that water contents between 10% and 12% allow germination to be maintained for a period of six to eight months for most species.

The thousand seed weight of *A. nahoumii* was 1.97 g (Table 1). Silva

(2006) verified that *Bromelia karatas* presented a thousand seeds weight of 47.80 g, in *Nidularium inncentii* and *N. procerum* the thousand seeds weight was 2.14 g and 1.99 g respectively (PEREIRA, 2009). The weight of one thousand seeds, which is generally used to calculate the sowing density and the weight of the work sample, is information that gives an idea of the quality of the seeds, as well as their state of maturity and sanity (BRASIL, 1992).

Table 1. Physical characteristics of the seeds of *Alcantarea nahoumii* (Leme) J, R. Grant. (Bromeliaceae).

Variables	Averages *
Water content (%)	14,71±0,34
Weight of one thousand seeds(g)	1,97±0,10

Means followed by standard deviations

Seeds of *A. nahoumii* germinate when kept for about 60 days, after sowing, in a BOD germination chamber between constant temperatures of 20 and 35 °C, under a 16 hour photoperiod. The highest germination averages were observed in geminate seeds at 20 ºC (51%) and at 25 ºC (45.5%). For these same temperatures, the first count and GVI variables had no significant effect between them, although they showed the highest averages. Seeds kept at 35 °C showed a low GVI (2.35), not exceeding 14.0% of total germination (Table 2).

Table 2 also shows that the mortality rate of A. Nahoumii seeds was considered low. At 20 °C there was a higher percentage of dead seeds, and at higher temperatures this response was reduced or even not detected (Table 2).

Table 2. Germination velocity index (GVI)and averages of firm and dead seeds of *Alcantarea nahoumii* (Leme) J. R. Grant (Bromeliaceae), at 60 days after germination.

Temperature	First count (18 days)	Germination	Seeds* IVGFirm (%)	Dead (%)
20°C	36,3 a	51,0 a	34,45 a46,75 c	2,25 a
25°C	39,0 a	45,5 b	35,75 a54,50 b	0,00 a
30°C	14,0 b	39,3 c	22,09 b59,25 b	1,50 a
35°C	0,0 c	19,5 d	2,35 c80,50 a	0,00 a

Means followed by different letters in the columns differ at 5% probability by Tukey's test

For Dyckia tuberosa Beer, the highest germinability and the shortest mean germination time were favored by temperatures between 30 and 35°C (VIEIRA et al., 2007). Pinheiro and Borghetti (2003) found high germinability for seeds of Aechmea nudicaulis and Streptocalyx floribundus Mez in Mart. at alternating temperatures of 20-30 °C . Similarly, Silva (2006) studying the germination of Bromelia karata found higher germination and GVI at alternating temperatures of 20-30 °C than at constant temperatures of 25 °C and 30 °C. According to Larcher (2006), for seeds to be able to germinate, their cardinal temperatures (minimum, optimum and maximum temperatures for development) must correspond to external conditions that ensure sufficiently rapid development for the young plants.

However, there were up to 80% of firm seeds that did not germinate at 35 °C, and it was also verified that with the increase of the temperature at which the tests were performed, there was an increase in the percentage of firm seeds (Table 2).

Anastácio and Santana (2010), studying the germination of *Ananas ananasoides,* related the slowness of the germination process and the high percentage of viability in relation to the final germination, to the poorly developed embryo. Pereira et at. (2009) studying the effect of temperatures on the germination behavior of four species of Bromeliaceae (*Alcantarea imperialis* (Carrière) Harms, *Pitcairnia flammea* Lindl., *Vriesea heterostachys* (Baker) L. B. Sm. and *Vriesea penduliflora* L. B. Sm.), verified that the seeds of the four species germinated between temperatures between 15 °C and 35 °C.

An exception was observed in *P. flammea* seeds, which germinated only between 15 °C and 25 °C. Similar results were obtained for seeds of *P. flammea* collected in São Paulo (MERCIER and GUERREIRO FILHO 1990). The ideal range for germination of the bromeliads studied was obtained between 20 °C and 25 °C, corroborating the information given by Mercier and Gu erreiro Filho (1990) for eight species of bromeliads occurring in the Atlantic Forest.

The first count of germination on the eighteenth day (Table 2) after the installation of the germination test was used as an evaluation of vigor. The first germination count is generally used to compare vigor between seed lots (KRZYZANOWSKI et al., 1999), however, here it was used to verify the effect of temperature on the initial performance of A. nahoumii seedlings. It was found that the temperature of 25°C showed the highest averages, allowing faster formation of the seedlings, while 35°C showed the lowest results, because it limited seed germination.

The final count was determined at 54 days, Figure 1, when the germination process stabilized and can be used to express the total percentage of germinated seeds. Based on this information, the germination test can be completed at 54 days for the species under study. According to Brazil, (1992) the germination test can be terminated even before the indicated time, when maximum germination has already been obtained.

Duarte (2007) verified better germination results in *Dyckia goehringii* Rauh & Gross at 30 °C, and recommended the evaluation of the first count on the seventh day after the beginning of the germination test. According to Barros et al. (2002), germination speed is reduced as seed deterioration progresses. Thus, samples that present higher germination values in the first count can be considered more vigorous.

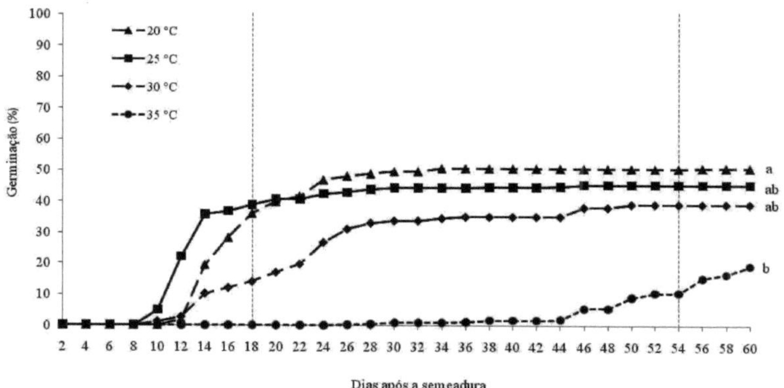

Figure 1. Germination behavior of *Alcantarea nahoumii* (Leme) J. R. Grant (Bromeliaceae) seeds over 60 days in different temperatures. Means followed by different letters, in the last evaluation, differ from each other at 5% probability by the Tukey test.

For seeds kept under the temperature of 20 ° C and 25 °C it took 24 days for the maximum germination percentage to be reached, and 48 days for seeds kept at 30 °C. For most of the temperatures tested, a relatively low maximum germination level was observed. For the temperature of 35 °C, a stabilization of germination was not detected until the end of the experiment (Figure 1).

There is often an ecological relationship between germination speed and climatic conditions. In species that under natural conditions germinate in the summer period, germination occurs very slowly at low temperatures, accelerating the process only after the substrate has reached more than 10 ºC (MOLLO, 2009). Thus, the synchronization is performed according to the most favorable season for the growth of young plants, improving their chances of survival and continued growth (LARCHER, 2006).

A. nahoumii is an endemic species of mountainous regions (MARTINEL LI et al., 2008), with average daily temperatures probably around 25 °C, but varying a lot during the same day (from 20 °C to 35 °C); therefore, the species seems to show some plasticity to temperature variations.

The vigor ofA. nahoumii seedlings was affected by temperature, with significantly higher differences for aerial part length, roots and total fresh mass, with the highest means at 25 °C and 30 °C (Table 3). Possibly the seedlings were more vigorous due to the acceleration in the respiratory metabolism of reserves (MELO et al., 2004) during the germination process.

Table 3. Mean aerial part length, root length, total fresh matter mass of *Alcantarea nahoumii* (Leme) J. R. Grant (Bromeliaceae), at 60 days after germination.

Temperature	Aerial Part (mm)	Root (mm)	Total fresh mass (mg)
20 °C	0,63 b	0,15 b	1,37 b
25 °C	12,12 a	9,75 a	33,17 a
30 °C	11,16 a	8,32 a	24,83 a
35 °C	0,00 b	0,00 b	0,00 b

Means followed by different letters in the columns differ at 5% probability by Tukey's test.

During the germination tests under 35 °C, the germination of seeds was observed (Table 2), however, there was no evolution in the development and growth of seedlings, (Ta bela 3), which presented initial development similar to the others, withering after a certain period. This raises the need for specific studies on the causes of this phenomenon.

Specific studies will be necessary for the identification of vigorous and normal seedlings in Bromeliaceae, characteristics that can guarantee their survival in natural and/or cultivation conditions, aiming to establish future laboratory routines. The cardinal temperatures and their relations with seedling vigor and development, as well as the time and conditions necessary for seedling storage should also be studied in order to estimate their longevity.

CONCLUSIONS

Alcantarea nahoumii seeds show restrictions for satisfactory germination under 35°C temperature.

To obtain more vigorous seedlings of *A. nahoumii*, seeds should be used at a temperature of 25°C.

The seed germination test can be evaluated at 18 days and 54 days after sowing.

The first count of seed germination can be used routinely to obtain preliminary information about seed lots of *A. nahoumii*.

BIBLIOGRAPHIC REFERENCES

ABDO, M. T. V. N.; PAULA, R. C. Temperatures for germination of capixingui (Crotonfloribundus Spreng - Euphorbiaceae) seeds. **Revista Brasileira de Sementes,** Pelotas, v.28, n.3, p. 135-140, 2006.

ANASTÁCIO, M. R.; SANTANA, D. G. Germination characteristics of seeds of *Ananas ananassoides* (Baker) L. B. Sm. (Bromeliaceae). **Acta Scientiarum. Biological Sciences.** Maringá, v. 32, n. 2, p. 195-200, 2010.

ANDRADE, A. C. S.; CUNHA, R.; SOUZA, A. F.; REIS, R. B.; ALMEIDA, K. J. Physiological and morphological aspects of seed viability of a neotropical savannah tree, *Eugenia dysenterica* DC. **Seed Science & Technology, n.** 3, p. 125-137, 2003.

BARROS, D. I.; NUNES, H. V.; DIAS, D. C. F. S.; BHERING, M. C. Comparação entre testes de vigor para avaliação da qualidade fisiológica de sementes de tomato. **Revista Brasileira de Sementes,** Pelotas, v. 24, n. 2, p.12-16, 2002.

BORGHETTI, F.; FERREIRA, A. G. Interpretação de resultados de germinação. In: FERREIRA, A. G., BORGHETTI, F., (Orgs.). **Germination** : do básico ao aplicado. Porto Alegre: Artmed, 2004. p. 209-222.

BRAZIL, Ministry of Agriculture and Agrarian Reform. **Regras para análise de sementes**. Brasília: SNDA/DNDV/CLAV, 1992. 365 p.

DUARTE, E. F. **Caracterização, qualidade fisiológica de sementes e crescimento inicial de *Dyckia goerhringii* Gross & Rauh, bromélia nativa do cerrado.** Thesis (Doctorate in Agronomy). Universidade Federal de Goiás. Goiás, 2007. 200 p.

FERREIRA, D.F. Analises estatísticas por meio do Sisvar para Windows versão 4.0. In: Reunião anual da Região Brasileira da soci edade Internacional de Biometria, 45, São Carlos, **Programas e resumos...** São Carlos: UFSCar, July 2000, p.255-258.

GOMES, V.; FERNANDES, G. W. Germination of *Baccharis dracunlifolia* D. C. (Asteraceae). **Acta Botânica Brasílica** , São Paulo, v. 16, n. 4, p. 421-427, 2002.

KRZYZANOWKI, F. C.; VIEIRA, R. D. D. Controlled deterioration. In: KRZYZANOWKI, F. C.; VIEIRA, R. D.; FRANÇA NETO, J. B. (Ed.). **Vigor of seeds: concepts and tests.** Londrina: ABRATES, 1999. p.61 -68.

LABOURIAL, L. G. **The Germination of Seeds**. Washington: General Secretariat of the Organization of American States, 1983. 173 p.

LARCHER, W. **Plant Ecophysiology**. RiMa, Sao Carlos, SP, 2006. 398p.

MAGUIRE, J. D. Speed of germination-aid in seletion and evaluation seedling emergence and vigor. **Crop Science**, Madison, v. 2, p. 176-177, 1962.

MARCOS-FILHO, J. **Fisiologia de sementes de plantas cultivadas**. Piracicaba: Fealq, 2005, 495 p.

MARTINELLI, G.; VIEIRA, C. M.; GONZALEZ, M.; LEITMAN, P.; PIRATININGA, A.; COSTA, A. F.; FORZZA, R. C. Bromeliaceae of the Brazilian Atlantic Forest: species list, distribution and conservation. **Rodriguésia**, v. 59 n.1, p. 209-258. 2008.

MELO, F. P. L.; AGUIAR NETO, A. V.; SIMABUKURO, E. A.; TABARELLI, M. Seedling recruitment and establishment. In: FE RREIRA, A. G. and BORGHETTI, F. (eds.). **Germination: do básico ao aplicado**. Porto Alegre, Artmed, p. 237-249, 2004.

MERCIER, H.; GUERREIRO FILHO, O. Propagacao sexuada de algumas bromélias nativas da Mata Atlântica: efeito da luz e da temperatura na germinação. **Hoehnea,** v.17, p. 19-26. 1990.

MOLLO, L. **Effect of temperature on growth, content and composition of non-structural carbohydrates of Alcantarea imperialis (Carrière) Harms (Bromeliaceae) plants cultivated *in vitro*.** Dissertation (Master in Plant Biodiversity and Environment) -- Botanical Institute of the State Secretariat of Environment, Sao Paulo, 2009. 90p.

PEREIRA, T. S. Bromelioideae (Bromeliaceae): morphology of post-seminal development of some species. **Arquivos do jardim Botânico do Rio de Janeiro**, Rio de Janeiro v. 29, p. 115-154, 1988.

PEREIRA, A. R.; ANDRADE, A. C. S. de; PEREIRA, T. S.; FORZZA, R. C.; RODRIGUES, A. S. Germinative behavior of epiphytic and ruicula species of Bromeliaceae from Ibitipoca State Park, Minas Gerais, Brazil. **Revista Brasileira de Botânica** v.32, n. 4, Oct./Dec. 2009

PINHEIRO , F.; BORGHETTI, F. Light and temperature requiriments for gerination of seeds of Aechmea nudicaulis (L.) Griesebach and Streptocalyx floribundus (Martius ex Schultes F.) Mez (Bromeliaceae). **Acta Botânica Brasílica** , São Paulo, v. 17, n. 1, p. 27-35, 2003.

POPINIGIS, F. **Fisiologia da semente**. Brasilia: AGIPLAN, 1985. 289p.

QUEIROZ, L. P.; SENA, T. S. N. and COSTA, M. J. L. S. Flora vascular da Serra da Jibóia, Santa Terezinha Bahia. In: O Campo Rupestre. **Sitientibus,** v. 15, p. 27-40, 1996.

SALISBURY, F. B.; ROSS, C. W. **Plant Physiology**. 4th ed. California; Wadsworth Publishing Company, 1991. 682p.

SILVA, E. E. **Frutiferas Nativos do Nordeste: qualidade fisiológi ca, morfologia e citogenética.** Dissertation (Master in Agronomy). Federal University of Paraíba. Paraíba, 2006. 110p.

SIQUEIRA FILHO, J. A.; TABARELLI, M. Bromeliads species of the Atlantic Forest of north-east Brazil: losses of critical populations of endemic species. **Oryx. Cambridge**, v. 40, n. 2, p. 1-7, 2006.

SMITH, L. B. ; DOWNS, R. J. Pitcairnoideae (Bromeliaceae).The New York Botanical Garden. **Fl. Neotrop. Monagr.** V. 14, n. 1, p. 1-658. New York, 1974.

VIEIRA, D. C. M.; SOCOLOWSKI, F.; TAKAKI, M. Germination of seeds of *Dyckia tuberosa (*Vell.) Beer(Bromeliaceae) under different temperatures in light and dark. **Revista Brasileira de Botânica** , v. 30, n. 2, p. 183-188, 2007.

WIELEWICKI, A. P.; LEONHARDT, C.; SCHLINDWEIN, G.; MEDEIROS, A. C. S. Proposta de padrões de germinação e teor de água pa ra sementes de algumas espécies florestais presentes na região sul do Brasil. **Revista Brasileira de Sementes,** v.28, n.3, p.191-197. 2006.

CONCLUDING REMARKS

The morphological and physiological analysis of the seeds of *Alcantarea nahoumii,* showed ecological adaptation for anemophilic dispersal, high germination capacity, however, there is some limitation in the process of seedling establishment, due to its slow development, which results in a species vulnerable to climatic adversities.

All the works defined methods of germination analysis for *A. nahoumii* seeds, characterizing normal seedlings, the moment of the first count and the final germination count, indicating the characteristics of the substrates and the temperatures that favor the best development of the seedlings.

Certainly, these studies will contribute to the expansion of the knowledge base about the Bromeliaceae family occurring in the Atlantic Forest, presenting a synthesis of information on part of the reproductive strategies undertaken by *Alcantarea nahoumii,* which are certainly the first steps for its dissemination in the technical-scientific environment, and may also contribute to its domestication and conservation. However, we emphasize the need for greater efforts and investments in research, awareness campaigns, and dissemination of the potential for using species of native Brazilian flora in a sustainable way, in order to promote the preservation of genetic heritage and biodiversity for future generations.

Printed by Books on Demand GmbH, Norderstedt / Germany